Cloud Certification Program for Game Changers

Learn. Practice. Get Certified.

ANGEL BERNIZ

International Institute of Executive Careers (IIEC)

Program Contents

Welcome to the Program

Thank you very much for purchasing this book.

First of all, let's connect on LinkedIn and join my **Game Changers** Newsletter on LinkedIn, so that you can get my insights, learnings, and exclusive contents to help you get career success:

https://www.linkedin.com/in/angelberniz

I would love you to share your experience with me after completing the program. You can message me via LinkedIn.

Okay, let's move on!

Certification Program

This book is the foundation for the **IIEC Cloud Certification Program™** available here:

https://thebestmba.org/program/cloud-certification

The Certification Program is available in the online program and will improve your job opportunities and certify your skills gained from this book.

The exams in the online Certification Program are based on the exam questions covered along this book. And once you sign up, you can take your online program whenever you want (no deadline limits). Just sign up today for your Certification Program and certify your skills.

Now, let's start enjoying this journey.

Preface

As we progress further into the 21st century, we witness a transformative era where cloud transformation is revolutionizing the way businesses operate and innovate. This paradigm shift in technology presents both unparalleled challenges and unique opportunities for those who aspire to lead the change and make a lasting impact. In this fast-paced and dynamic landscape, it is crucial for these game changers to adopt a strategic, secure, and flexible approach to navigate the world of cloud transformation.

The purpose of this program is to provide a practical and accessible resource for game changers, empowering you to harness the potential of cloud transformation and lead the way in shaping the future of businesses. Drawing on insights from a diverse array of disciplines, including computer science, information technology, business management, and security, this program seeks to foster a comprehensive and interdisciplinary understanding of cloud transformation and its implications for various industries.

Throughout the program, we will lay a foundation for understanding the key principles and concepts that underpin cloud technologies and discuss the security considerations that arise as we integrate these advancements into the business landscape. The program offers strategies, best practices, and actionable insights designed to empower game changers in their pursuit of innovation and positive change.

As we delve deeper into the future of business in the cloud transformation era, we will examine the potential impact of cloud technologies on various sectors, highlighting both the transformative opportunities and the challenges that lie ahead.

Throughout the program, we will underscore the need for responsible innovation, interdisciplinary collaboration, stakeholder engagement, and global cooperation to navigate the challenges and seize the opportunities presented by the cloud transformation era. By taking an active role in shaping the future of businesses, game changers can ensure that cloud technologies are developed and deployed in ways that align with our collective values, security principles, and organizational goals.

In offering this program, it is my hope that participants will come away with a deeper understanding of the cloud transformation era and its implications for the future of businesses, as well as a sense of optimism, inspiration, and empowerment. As game changers, you have a unique opportunity to shape the course of cloud development in ways that reflect our shared aspirations for a more efficient, scalable, and sustainable world. Together, let us embrace the challenges and opportunities of the cloud transformation revolution and chart a bright future for businesses and technology, hand in hand.

Acknowledgments

My notes at the university were: "Early adopters are the wealthiest people on earth."

Little did I know how those words would shape my path. As I reflect on my journey, I am overwhelmed with gratitude for the countless individuals who accompanied, guided, and inspired me along the way.

I owe a profound debt of gratitude to my mentors, colleagues, and collaborators in executive education and the technology industry. Their unwavering support and willingness to share their knowledge and perspectives have been invaluable. Their guidance has shaped my career and fueled my passion.

To my students at the International Institute of Executive Careers (IIEC), you have been my constant inspiration. Your curiosity and dedication have made me a better educator and challenged me to refine my ideas. Your insights have significantly contributed to this book.

I also acknowledge the remarkable researchers, professionals, and practitioners who have paved the way for our understanding of the potential of emerging technologies. Their visionary thinking and tireless pursuit of knowledge continue to inspire individuals worldwide, including myself.

I am deeply grateful for the unwavering love, support, and understanding of my friends and family. Your belief in me and willingness to engage with my ideas have given me strength and resilience. This book is dedicated to you with love and gratitude.

As I conclude, I am reminded of the interconnected relationships, experiences, and ideas that have shaped this book. To all who have touched my life, big or small, thank you. Your presence has been truly meaningful. Together, let us continue exploring the evolving landscape of technology, guided by the belief that early adopters can shape the future.

About the Author

Angel Berniz is a seasoned Business and Technology executive, strategist, and advisor with over 20 years of experience in the technology industry. He has held various leadership positions in several companies and has worked with clients in different industries, helping them navigate cloud transformation, innovation, sustainability, emerging technologies, and cybersecurity.

Angel Berniz is also a recognized keynote speaker and author on topics related to business technologies, leadership, and innovation. He has been featured in several conferences, seminars, and publications. He is also a professor and lecturer in international executive programs of several Business Schools.

You can explore Angel Berniz's extensive collection of books available on Amazon:

https://www.amazon.com/author/angelberniz .

Module 1. Setting Up a Cloud Environment

Introduction to Setting Up a Cloud Environment

Welcome to the first module of our practical Cloud Transformation program! In this module, we will introduce you to the crucial first step in any cloud transformation journey - setting up a robust cloud environment. A well-designed and secure cloud environment is the foundation for a successful digital transformation, and it paves the way for seamless deployment, management, and optimization of your applications and services.

Choosing the right cloud service provider is essential to the success of your cloud transformation. There are several key players in the market, each with its unique offerings, pricing models, and service features. Amazon Web Services (AWS), Microsoft Azure, and Google Cloud Platform (GCP) are the three leading cloud service providers, with a vast array of services catering to various industry needs. To help you make an informed decision, we will walk you through a comparison of these providers, highlighting their strengths and weaknesses. We will also touch upon other alternatives such as IBM Cloud and Oracle Cloud, which might be more suitable for specific use cases.

Once you have selected a cloud service provider, you need to create and manage your cloud account. This process involves setting up billing, organizing resources, and applying best practices for access management. We will guide you through the process of creating an account and ensuring that it is well-structured for future growth and expansion.

One of the primary practical examples we will delve into during this module is setting up a virtual machine (VM) in your chosen cloud provider. VMs are the backbone of many cloud-based applications, providing a scalable and flexible infrastructure that can be adjusted according to your business requirements. We will demonstrate how to create, configure, and manage a VM, ensuring that you have a firm understanding of this essential cloud resource.

Throughout this module, you will gain hands-on experience with the tools and techniques necessary for setting up a secure and efficient cloud environment. By the end of this module, you will be well-equipped with the knowledge and skills required to confidently embark on your cloud transformation journey.

Choosing a Cloud Service Provider (AWS, Azure, GCP)

In this section, we will focus on the critical decision-making process of selecting the right cloud service provider for your organization's cloud transformation journey. With a plethora of options available in the market, it is essential to understand the unique offerings and strengths of each provider to ensure that your chosen platform aligns with your business needs and objectives.

Amazon Web Services (AWS), Microsoft Azure, and Google Cloud Platform (GCP) are the three leading cloud service providers, dominating the cloud computing market. Each provider brings a unique set of services, features, and capabilities to the table, catering to various industry requirements and business use cases. To help you make an informed decision, we will provide an overview of these providers, highlighting their differences and similarities, and guiding you through the factors that should be considered when choosing a cloud service provider.

Throughout this section, we will cover real-world examples and case studies, demonstrating how organizations from different industries have selected and leveraged these cloud platforms to achieve their business goals. By examining these practical examples, you will gain a better understanding of the decision-making process and the factors that influence the choice of a cloud service provider.

By the end of this section, you will have a clear understanding of the critical aspects of AWS, Azure, and GCP, and be equipped with the knowledge necessary to make an informed decision on the right cloud service provider for your organization's cloud transformation journey.

Understanding the Market Landscape

The cloud computing market is dominated by three major players: Amazon Web Services (AWS), Microsoft Azure, and Google Cloud Platform (GCP). Each provider has carved out a niche in the industry, serving various business requirements and use cases. In this section, we will explore the market landscape of these providers, providing you with insights into their market share, strengths, weaknesses, and how they cater to different industry needs.

Amazon Web Services (AWS) is the current market leader, holding the largest share of the cloud computing industry. Launched in 2006, AWS has a head start over its competitors, resulting in a more extensive range of services, more mature infrastructure, and a larger customer base. AWS is known for its scalability, reliability, and performance, making it a popular choice for organizations of all sizes. Its comprehensive service offerings include computing power, storage, databases, machine learning, analytics, and more. Examples of companies leveraging AWS for their cloud needs include Netflix, Airbnb, and Adobe.

Microsoft Azure, launched in 2010, holds the second-largest market share. As a part of the Microsoft ecosystem, Azure is particularly attractive to organizations that rely heavily on Microsoft products and services. Azure provides seamless integration with tools such as Office 365, Power BI, and Dynamics 365. Microsoft Azure offers a wide range of services, including virtual machines, storage, databases, and machine learning, catering to various industries, including healthcare, finance, and government. Companies that have chosen Azure for their cloud transformation include BMW, Samsung, and GE Healthcare.

Google Cloud Platform (GCP), although a relatively late entrant in the market (launched in 2011), has gained significant traction and holds the third-largest market share. GCP stands out for its cutting-edge technology, data analytics, and machine learning capabilities, making it a popular choice for organizations focused on innovation and data-driven decision-making. GCP's service offerings include computing power, storage, databases, Big Data and analytics, and artificial intelligence services. Examples of companies using GCP for their cloud needs are Spotify, Snap Inc., and HSBC.

Each cloud provider has its unique set of strengths and weaknesses, catering to different industry needs. By understanding the market landscape, you can make informed decisions about which provider best aligns with your organization's requirements, resources, and business goals.

Comparing Features and Services

Comparing Features and Services: Evaluate the core services provided by AWS, Azure, and GCP, including compute, storage, networking, and security. Understand the differences in their service offerings, performance, and ease of use.

When selecting a cloud service provider, it is essential to compare the features and services offered by each provider to determine which one best meets your organization's needs. In this section, we will evaluate the core services provided by AWS, Azure, and GCP, focusing on compute, storage, networking, and security. By understanding the differences in their service offerings, performance, and ease of use, you can make an informed decision about which provider is the best fit for your cloud transformation journey.

- **Compute Services:** AWS offers a range of compute services, including Elastic Compute Cloud (EC2) for virtual machines, Lambda for serverless computing, and Elastic Beanstalk for platform-as-a-service (PaaS). Azure provides similar services, such as Azure Virtual Machines, Azure Functions, and Azure App Service. GCP's offerings include Compute Engine, Cloud Functions, and App Engine. While all three providers offer comparable compute services, the performance, pricing, and ease of use may vary.

For example, AWS EC2 instances offer a wide variety of configurations, allowing you to tailor resources to your specific needs. Azure Virtual Machines provide seamless integration with other Microsoft services and tools, simplifying the deployment and management process. GCP's Compute Engine is known for its fast boot times and efficient performance, making it attractive for compute-intensive workloads.

- **Storage Services:** All three providers offer a range of storage services, including object storage, block storage, and file storage. AWS provides Amazon Simple Storage Service (S3) for object storage, Elastic Block Store (EBS) for block storage, and Amazon Elastic File System (EFS) for file storage. Azure offers Azure Blob Storage, Azure Disk Storage, and Azure Files, respectively. GCP provides Cloud Storage, Persistent Disk, and Filestore.

  While all three providers offer robust and scalable storage solutions, there may be differences in pricing, performance, and available features. For instance, AWS S3 is known for its durability and scalability, while Azure Blob Storage offers seamless integration with the Microsoft ecosystem. GCP's Cloud Storage is recognized for its cost-effectiveness and performance, especially for large-scale data storage.

- **Networking Services:** AWS, Azure, and GCP all provide comprehensive networking services, including virtual networks, load balancing, content delivery networks (CDNs), and Domain Name System (DNS) management. AWS offers Amazon Virtual Private Cloud (VPC), Elastic Load Balancing, Amazon CloudFront, and Amazon Route 53. Azure provides Azure Virtual Network, Azure Load Balancer, Azure CDN, and Azure DNS. GCP offers Google Cloud VPC, Cloud Load Balancing, Cloud CDN, and Cloud DNS.

The choice of networking services may depend on your specific requirements, such as global reach, performance, or ease of configuration. For example, Amazon CloudFront has a vast global network, ensuring low-latency content delivery. Azure Virtual Network allows for easy integration with on-premises networks using Azure ExpressRoute, while GCP's Cloud Load Balancing offers advanced traffic management features, such as global load balancing across multiple regions.

- **Security Services:** Security is a top priority for any organization moving to the cloud. AWS, Azure, and GCP all provide robust security services, including identity and access management (IAM), data encryption, and compliance certifications. AWS offers AWS IAM, AWS Key Management Service (KMS), and AWS Security Hub. Azure provides Azure Active Directory, Azure Key Vault, and Azure Security Center. GCP offers Cloud Identity, Cloud Key Management Service, and Google Cloud Security Command Center.

  Each provider offers comprehensive security services, but your organization's existing security policies and requirements may influence your choice. For example, AWS IAM is known for its granularity and flexibility in managing user access, while Azure Active Directory offers seamless integration with existing Microsoft-based identity systems. GCP's Cloud Identity focuses on simplicity and ease of use, making it attractive for organizations that prefer a more straightforward approach to identity management.

In summary, AWS, Azure, and GCP each provide a wide array of features and services across compute, storage, networking, and security. While their offerings are generally comparable, there may be differences in performance, ease of use, and integration with existing tools and systems. To choose the right cloud service provider, evaluate their features and services based on your organization's specific needs, resources, and goals. By considering real-world examples and case studies, you can gain a deeper understanding of how each provider's offerings can be leveraged to achieve success in your cloud transformation journey.

Assessing Pricing and Cost Models

Understanding the pricing and cost models of cloud service providers is crucial for managing your organization's cloud expenditure and optimizing resources. In this section, we will compare the pricing structures of AWS, Azure, and GCP, focusing on pay-as-you-go, reserved instances, and spot pricing. By examining these pricing models and understanding the factors that contribute to the overall cost of running applications and services, you can make informed decisions about which provider best aligns with your organization's financial goals.

- **Pay-as-you-go:** The pay-as-you-go model is the most common pricing model in cloud computing, with all three providers offering this option. With pay-as-you-go, you only pay for the resources you consume without any upfront commitment. This model provides flexibility and allows you to scale resources as needed, but it may not be the most cost-effective option for long-term, predictable workloads.

  For example, AWS charges per hour or per second for EC2 instances, depending on the instance type. Azure bills per minute for virtual machines, while GCP charges per second with a minimum of one minute. To optimize costs in this model, you should closely monitor resource usage and adjust resources as needed.

- **Reserved Instances:** Reserved instances (RIs) allow you to reserve compute capacity for a specific period, typically one or three years, in exchange for a discounted rate compared to the pay-as-you-go pricing. RIs are best suited for predictable workloads with stable resource requirements.

AWS offers Reserved Instances for EC2 instances, providing discounts of up to 72% compared to on-demand pricing. Azure provides a similar offering called Reserved Virtual Machine Instances, with discounts of up to 72%. GCP offers a unique approach called Committed Use Contracts, which provide discounts of up to 57% for committing to use specific resources for one or three years.

- **Spot Pricing:** Spot pricing allows you to bid on unused cloud resources at significantly discounted rates, often up to 90% off the pay-as-you-go pricing. This model is best suited for short-term, flexible workloads that can tolerate interruptions.

  AWS offers Spot Instances, where you bid on unused EC2 capacity. Azure provides a similar service called Azure Spot Virtual Machines. GCP offers Preemptible VMs, which are short-lived instances available at a fixed discounted price.

It's essential to consider factors such as data transfer, storage, and additional services when assessing the overall cost of running applications and services in a cloud environment. For instance, data transfer costs can vary between providers and may impact your total expenditure.

In summary, understanding the pricing structures and cost models of AWS, Azure, and GCP is crucial for managing cloud expenses and optimizing resources. By comparing pay-as-you-go, reserved instances, and spot pricing models, you can make informed decisions about which provider best suits your organization's financial goals and workload requirements.

Analyzing Integration and Compatibility

Ensuring compatibility and seamless integration between your existing tools, technologies, and platforms with your chosen cloud provider is essential for a smooth cloud transformation journey. In this section, we will explore the compatibility of AWS, Azure, and GCP with your current ecosystem and evaluate their support for popular programming languages, frameworks, and third-party integrations.

- **Programming Languages:** All three major cloud providers support a wide range of popular programming languages, making it easier for developers to work with the platforms. AWS, Azure, and GCP support languages such as Python, Java, JavaScript (Node.js), C#, Go, and Ruby, among others. However, the level of support and available features may vary between providers.

  For example, AWS Lambda, a serverless compute service, supports Python, Node.js, Java, C#, and Go, making it versatile for different development teams. Azure Functions, a similar serverless offering from Microsoft, provides additional support for F# and PowerShell. GCP's Cloud Functions also support a variety of languages, including Python, Node.js, Go, and Java.

- **Frameworks:** AWS, Azure, and GCP all offer support for popular frameworks, such as TensorFlow, PyTorch, and Apache MXNet for machine learning, as well as Django, Flask, and Express.js for web development. The choice of provider may depend on the specific features and performance requirements for your project.

  For instance, GCP is known for its deep integration with TensorFlow, as both are products of Google. This integration provides performance optimizations and an extensive suite of tools for developing and deploying TensorFlow models. Azure, on the other hand, offers seamless integration with Microsoft's ML.NET framework, making it an attractive option for organizations heavily invested in the Microsoft ecosystem.

- **Third-party Integrations:** AWS, Azure, and GCP all provide integration with various third-party tools and services, enabling you to extend the capabilities of your cloud environment. These integrations may include monitoring and logging tools, such as Datadog or Splunk, as well as continuous integration and deployment (CI/CD) platforms like Jenkins, CircleCI, or GitLab.

  For example, AWS offers integration with popular third-party services through AWS Marketplace, an online store that allows customers to find, buy, and deploy software and services for their AWS environments. Similarly, Azure provides integration with various third-party tools through Azure Marketplace, while GCP offers integrations via Google Cloud Marketplace.

In summary, analyzing the integration and compatibility of AWS, Azure, and GCP with your existing tools, technologies, and platforms is essential for ensuring a successful cloud transformation. By evaluating their support for popular programming languages, frameworks, and third-party integrations, you can select a cloud provider that best aligns with your organization's existing ecosystem and technical requirements. This compatibility will result in a smoother transition and allow you to leverage the full potential of your chosen cloud platform.

Evaluating Support and Documentation

Reliable customer support and comprehensive documentation are crucial for the success of your cloud transformation journey. In this section, we will evaluate the quality of customer support and available documentation for AWS, Azure, and GCP, focusing on factors such as response times, community support, training resources, and the availability of professional services.

- **Response Times:** When facing technical issues or requiring assistance, timely and effective support from your cloud provider is essential. AWS, Azure, and GCP all offer various support plans, with different levels of responsiveness and features depending on the chosen plan. Free plans typically provide slower response times and limited access to support resources, while premium plans offer faster response times and additional support features.

For example, AWS offers four support plans: Basic, Developer, Business, and Enterprise. Response times for the premium plans range from 12 hours for the Developer plan to 15 minutes for business-critical system down issues in the Enterprise plan. Azure provides similar support plans, with response times ranging from 8 hours to 15 minutes, depending on the severity of the issue and the plan selected. GCP also offers multiple support plans, with response times ranging from 4 hours to 15 minutes for critical issues.

- **Community Support:** All three providers have active and extensive user communities that can offer valuable insights, tips, and troubleshooting assistance. AWS, Azure, and GCP all have community forums, Stack Overflow channels, and social media groups where users can ask questions, share knowledge, and collaborate. These communities can be invaluable resources when facing challenges or seeking best practices in implementing cloud services.

- **Training Resources:** AWS, Azure, and GCP all provide extensive training resources, including online documentation, whitepapers, video tutorials, and webinars. These resources cover a wide range of topics, from getting started with the platform to advanced techniques for optimizing performance and security.

  For example, AWS offers AWS Training and Certification, a comprehensive set of learning resources and certifications for different roles and skill levels. Azure provides Microsoft Learn, a similar platform that offers learning paths and certifications tailored to Azure services. GCP's Google Cloud Training offers a wide range of courses and certifications, covering topics such as data engineering, machine learning, and security.

- **Professional Services:** In addition to support and documentation, AWS, Azure, and GCP all offer professional services to help organizations navigate their cloud transformation journeys. These services can provide guidance and hands-on assistance with architecture design, implementation, and optimization.

  AWS Professional Services, Azure Professional Services, and Google Cloud Professional Services all have teams of experts who can work closely with your organization to ensure the success of your cloud initiatives. Additionally, all three providers have a network of partners that can offer specialized support and consulting services tailored to your specific needs and industry.

In summary, evaluating the quality of customer support and available documentation for AWS, Azure, and GCP is essential for ensuring a successful cloud transformation. By considering factors such as response times, community support, training resources, and the availability of professional services, you can choose a provider that offers the right level of support and resources for your organization's needs. This strong support foundation will contribute to a smooth and efficient cloud transformation process.

Creating and Managing Cloud Accounts

Properly setting up and managing your cloud accounts is a crucial aspect of a successful cloud transformation journey. In this section, we will cover the essential steps and best practices for creating, organizing, and maintaining cloud accounts with AWS, Azure, or GCP. Understanding these fundamentals will help you ensure that your organization's cloud environment is secure, well-organized, and cost-effective.

We will explore the process of signing up for an account with your chosen cloud provider and the necessary steps for verification, authentication, and initial setup. You will learn about the various approaches to organizing resources within your cloud account and how to efficiently manage those resources. This includes utilizing projects, resource groups, or accounts and implementing tagging strategies to streamline resource management.

Next, we will delve into access control and user management, discussing how to leverage Identity and Access Management (IAM) features, roles, and policies to manage user access and permissions within your cloud environment. Security best practices, such as enabling multi-factor authentication (MFA), setting up billing alerts, and regularly auditing permissions and resource usage, will also be highlighted to help protect your cloud account and ensure a secure cloud infrastructure.

Finally, we will examine tools and techniques for monitoring and optimizing costs within your cloud account. This includes understanding the cost management features provided by AWS, Azure, and GCP, setting up budgets, allocating costs, and using usage analytics to track and optimize your organization's cloud expenditure.

By the end of this section, you will have a comprehensive understanding of the key aspects involved in creating and managing cloud accounts with AWS, Azure, or GCP. This knowledge will empower you to build and maintain a secure, organized, and cost-effective cloud environment that supports your organization's cloud transformation journey.

Account Setup and Onboarding

Setting up a cloud account is the first step in your organization's cloud transformation journey. In this section, we will explore the process of signing up for an account with AWS, Azure, or GCP and guide you through the necessary steps for verification, authentication, and initial setup.

- **AWS Account Setup:** To create an AWS account, you will need to provide an email address, create a password, and enter your contact information. During the signup process, you will also be prompted to enter your credit card details for billing purposes. AWS offers a free tier, which includes a limited amount of resources that can be used at no cost for the first 12 months. After completing these steps, AWS will require you to verify your phone number and choose a support plan. Once your account is set up, you can access the AWS Management Console and start exploring AWS services.

- **Azure Account Setup:** Signing up for an Azure account requires a Microsoft account, which can be created with an existing email address. During the registration process, you will need to provide your contact information and credit card details, as Azure also offers a free tier with a limited amount of resources for the first 12 months. You will also need to verify your phone number and agree to the subscription agreement. After completing these steps, you can access the Azure portal to begin using Azure services.

- **GCP Account Setup:** To create a GCP account, you will need a Google account, which can be created with an existing email address. During the signup process, you will be asked to provide your contact information, credit card details, and agree to the terms of service. GCP offers a free tier, which includes a limited amount of resources that can be used at no cost for the first 12 months or until the credit is exhausted. Once your account is set up, you can access the Google Cloud Console to start using GCP services.

In all cases, it's crucial to follow best practices for securing your account, such as enabling multi-factor authentication (MFA) and setting up strong, unique passwords for each user. Additionally, you should familiarize yourself with the management console or portal of your chosen provider, as this will be the primary interface for managing resources, monitoring usage, and configuring services.

In summary, understanding the process of signing up for an account with AWS, Azure, or GCP, including the necessary steps for verification, authentication, and initial setup, will lay the foundation for your organization's cloud transformation journey. By following best practices and familiarizing yourself with the management interface, you can efficiently navigate and manage your cloud environment.

Organizational Structure and Resource Management

Organizing resources effectively in your cloud account is crucial for efficient management, cost control, and maintaining a clear overview of your infrastructure. In this section, we will discuss best practices for organizing resources within AWS, Azure, and GCP, covering approaches such as projects, resource groups, or accounts, and implementing tagging strategies for better resource management.

- **AWS:** In AWS, you can create multiple accounts within an organization to separate resources and environments, such as development, staging, and production. AWS Organizations allows you to manage these accounts centrally and apply policies and permissions across all accounts. Resources within an account can be further organized using resource groups, which enable you to group resources that share common attributes.

Tagging is a powerful strategy to categorize resources in AWS based on user-defined key-value pairs. Tags can help you quickly identify resources, automate tasks, track costs, and enforce governance. For example, you can tag resources with environment ("prod", "dev"), project ("ProjectA", "ProjectB"), or owner ("John", "Jane") to make managing resources more efficient.

- **Azure:** In Azure, resources can be organized using resource groups, which are logical containers for resources deployed within an Azure subscription. Resource groups help you manage and monitor resources collectively, apply consistent policies, and provide access control. It is essential to create resource groups based on logical relationships between resources, such as by environment, project, or application.

  Azure also supports tagging resources with user-defined key-value pairs to categorize and manage resources more efficiently. Tags can be used for tracking costs, automating tasks, and enforcing governance, similar to AWS.

- **GCP:** In GCP, resources are organized using projects, which are the primary organizational unit for managing resources. Projects allow you to group resources that belong to a particular application, environment, or team. You can manage multiple projects under an organization, which enables centralized management of policies, permissions, and billing.

  Tagging resources in GCP is accomplished through labels, which are key-value pairs that can be attached to resources. Labels help you categorize resources, track costs, automate tasks, and enforce governance, similar to AWS and Azure.

In all three providers, it is essential to establish a consistent naming and tagging convention that is easy to understand and maintain. By doing so, you can quickly identify resources, automate tasks, track costs, and enforce governance.

In summary, understanding best practices for organizing resources within your cloud account, such as using projects, resource groups, or accounts, and implementing tagging strategies, will significantly improve the efficiency of managing your cloud environment. Adopting these practices will ensure a well-organized, transparent, and easily manageable infrastructure, supporting your organization's cloud transformation journey.

Access Control and User Management

Effectively managing user access and permissions is a critical aspect of maintaining a secure and well-governed cloud environment. In this section, we will discuss the different approaches for managing user access and permissions within AWS, Azure, and GCP, leveraging the Identity and Access Management (IAM) features, roles, and policies provided by each cloud provider.

- **AWS Identity and Access Management (IAM):** AWS IAM enables you to manage access to AWS services and resources securely. With IAM, you can create users, groups, and roles, and use permissions to allow or deny their access to resources. IAM supports fine-grained access control using policy statements that define actions, resources, and conditions.

  A practical example would be creating a group for your development team and attaching a policy that allows access to specific AWS services, such as EC2 and S3, only within the development environment.

- **Azure Active Directory (Azure AD) and Role-Based Access Control (RBAC):** Azure AD is a cloud-based identity and access management service that allows you to manage users, groups, and access to resources within your Azure environment. Azure RBAC is built on top of Azure AD and provides a way to grant permissions to resources using roles. Roles can be assigned to users, groups, or service principals at different scopes, such as a subscription, resource group, or individual resource.

  For instance, you can create a custom role for your operations team that grants them read-only access to resources in the production environment, ensuring they can monitor but not modify the resources.

- **Google Cloud Identity and Access Management (Cloud IAM):** Cloud IAM in GCP allows you to manage access control by defining who (identity) has what access (role) to which resource. In Cloud IAM, you can create and manage users, groups, and service accounts, and assign roles that determine the level of access to resources. Cloud IAM supports predefined roles as well as custom roles to fit your organization's specific needs.

  For example, you can grant a group of developers the "Compute Instance Admin" role, which allows them to manage virtual machines in a specific GCP project.

In all three providers, it is essential to follow the principle of least privilege when granting access, meaning users should only have the minimum level of access necessary to perform their tasks. Additionally, you should regularly audit user access and permissions to ensure they remain up-to-date and secure.

In summary, understanding and leveraging the Identity and Access Management (IAM) features, roles, and policies provided by AWS, Azure, and GCP will enable you to manage user access and permissions effectively within your cloud environment. By following best practices and maintaining a well-governed access control strategy, you can significantly enhance the security and manageability of your organization's cloud infrastructure.

Security Best Practices

Ensuring the security of your cloud account is a top priority when managing a cloud environment. In this section, we will discuss essential security measures that will help you protect your cloud account and safeguard your organization's data and infrastructure. These practices include enabling multi-factor authentication (MFA), setting up billing alerts, and regularly auditing permissions and resource usage.

- **Multi-Factor Authentication (MFA):** MFA is a security measure that requires users to provide two or more forms of identification to access their cloud account. This helps protect your account from unauthorized access, even if a password is compromised. All three major cloud providers (AWS, Azure, and GCP) support MFA, and it is highly recommended to enable MFA for all users, especially those with administrative privileges.

  For example, in AWS, you can enable MFA for IAM users by navigating to the IAM console, selecting the user, and then enabling MFA under the "Security credentials" tab. Users will then be prompted to provide a verification code from an authenticator app or hardware device when signing in.

- **Billing Alerts:** Setting up billing alerts is essential for monitoring and controlling costs within your cloud environment. Billing alerts can notify you when your cloud expenses reach a specific threshold, helping you prevent unexpected charges and manage your budget more effectively.

  In AWS, you can create billing alerts using Amazon CloudWatch, which allows you to set up alarms based on your estimated charges. Similarly, in Azure, you can use Azure Cost Management's budget feature to create alerts based on your spending patterns. In GCP, you can set up budget alerts using the Google Cloud Console, which will notify you when your spending reaches a specified amount.

- **Regular Auditing of Permissions and Resource Usage:** Regularly reviewing and auditing user permissions and resource usage is crucial for maintaining a secure and well-organized cloud environment. This practice helps you identify any excessive or unnecessary permissions, as well as unused or underutilized resources that could be potential security risks or contribute to increased costs.

  AWS, Azure, and GCP all provide tools and features that help you audit and monitor your environment. For example, AWS Trusted Advisor, Azure Advisor, and Google Cloud Security Command Center can provide recommendations and insights to help you optimize your security posture, resource usage, and cost management.

Implementing these security best practices will allow you significantly enhance the protection of your cloud account and maintain a secure, well-governed infrastructure. Ensuring the security of your cloud environment is a continuous process that requires ongoing monitoring, review, and adaptation to stay ahead of potential threats and vulnerabilities.

Cost Management and Optimization

Effectively managing costs is a vital aspect of operating a cloud environment. In this section, we will discuss tools and techniques for monitoring and optimizing costs within your cloud account, covering budgeting, cost allocation, and usage analytics provided by AWS, Azure, and GCP.

Budgeting:

Setting budgets allows you to plan and control your cloud spending, ensuring you stay within your financial constraints. All three major cloud providers offer budgeting tools to help you track and manage your costs.

- **AWS:** AWS Budgets allows you to set custom cost and usage budgets that alert you when your spending or usage exceeds the defined thresholds. You can create budgets based on various dimensions, such as service, linked account, or tag.
- **Azure:** Azure Cost Management and Billing offer budgeting features that enable you to create cost and usage budgets for your subscriptions, resource groups, or management groups. You can set up email alerts when your spending approaches or exceeds the budget thresholds.
- **GCP:** Google Cloud Platform provides budgeting features through the Google Cloud Console, which allow you to create budgets for your GCP projects or billing accounts. You can set up email notifications when your spending reaches specified percentages of the budget.

Cost Allocation:

Proper cost allocation helps you understand the costs associated with specific resources, projects, or business units within your organization. Implementing tagging strategies and using cost allocation tools provided by cloud providers can streamline this process.

- **AWS:** AWS Cost Explorer and Cost Allocation Tags enable you to categorize resources and view cost data based on your custom tags. This allows you to allocate costs to specific projects, teams, or environments for accurate cost reporting.
- **Azure:** Azure Cost Management provides cost allocation features through the use of tags and allows you to create

custom reports and dashboards that display cost data based on your tagging structure.

- **GCP:** Google Cloud Platform offers cost allocation features through the use of labels and the Google Cloud Billing Report, which allows you to view and analyze cost data based on your custom labels.

Usage Analytics:

Analyzing your resource usage helps you identify opportunities for cost optimization, such as resizing instances, deleting unused resources, or leveraging discounts and savings plans.

- **AWS:** AWS Cost Explorer provides insights into your resource usage and spending patterns, allowing you to identify cost-saving opportunities. Additionally, AWS Trusted Advisor can provide recommendations for cost optimization based on your usage patterns.

- **Azure:** Azure Advisor offers personalized cost-saving recommendations based on your resource usage, helping you optimize your Azure environment. Azure Cost Management also provides detailed reports and analytics that enable you to analyze your spending and usage.

- **GCP:** Google Cloud Platform provides cost and usage analytics through the Google Cloud Console, which allows you to analyze your spending patterns and identify potential savings. Google Cloud also offers recommendations for cost optimization based on your usage patterns.

Leveraging these tools and techniques will allow you effectively monitor and optimize costs within your cloud account. Adopting cost management best practices will enable you to run your cloud environment more efficiently and reduce overall expenses while still meeting your organization's performance and reliability requirements.

Practical: Setting Up a Virtual Machine (VM)

Setting up a virtual machine (VM) is a foundational skill when working with cloud environments. In this practical section, we will guide you through the process of creating, configuring, and managing a virtual machine on the cloud platform of your choice - AWS, Azure, or GCP. Virtual machines allow you to run applications and services on-demand, providing you with the flexibility to scale and optimize resources as needed.

Throughout this section, we will present detailed information and practical examples to help you get hands-on experience with creating and managing virtual machines in the cloud. By the end of this module, you will be familiar with the steps required to set up a VM, configure its settings, deploy and access it, and eventually clean up and terminate the instance when it's no longer needed.

Whether you are working with Amazon EC2 instances, Azure Virtual Machines, or Google Compute Engine instances, this practical guide will provide you with the necessary skills to confidently manage virtual machines in any cloud environment. This hands-on experience will not only reinforce your understanding of cloud concepts but also empower you to apply these skills in real-world situations, helping you optimize resource usage and enhance the performance of your cloud-based applications and services.

Selecting the Appropriate VM Instance Type

Choosing the right virtual machine instance type is crucial for ensuring optimal performance and cost efficiency for your cloud-based applications. Each cloud provider offers a variety of instance types, optimized for different use cases and workloads, such as compute-intensive applications, memory-intensive applications, or storage-intensive applications.

- **AWS EC2:** Amazon EC2 offers a wide range of instance types, grouped into categories like general-purpose, compute-optimized, memory-optimized, and storage-optimized instances. To select the appropriate instance type, consider factors such as the required number of vCPUs, memory, and storage for your application.

  Example: If your application requires 4 vCPUs and 16 GB of memory, you can choose a general-purpose instance like the m5.xlarge, which provides 4 vCPUs and 16 GB of memory.

- **Azure VM:** Azure Virtual Machines also offer a variety of instance types, categorized as general-purpose, compute-optimized, memory-optimized, storage-optimized, and GPU instances. Similar to AWS, consider the required vCPUs, memory, and storage when selecting an Azure VM instance type.

  Example: If your application needs 2 vCPUs and 8 GB of memory, you can choose a general-purpose instance like the D2s_v3, which offers 2 vCPUs and 8 GB of memory.

- **GCP Compute Engine:** Google Compute Engine provides a range of instance types, grouped into categories like general-purpose, compute-optimized, memory-optimized, and storage-optimized instances. As with AWS and Azure, consider the required vCPUs, memory, and storage when selecting a GCP Compute Engine instance type.

  Example: If your application requires 8 vCPUs and 32 GB of memory, you can choose a general-purpose instance like the n2-standard-8, which provides 8 vCPUs and 32 GB of memory.

When selecting an instance type, it's essential to consider your application's specific resource requirements to ensure optimal performance and cost efficiency. Additionally, it's a good practice to start with a smaller instance type and monitor your application's performance, scaling up as needed. This approach helps you avoid over-provisioning resources and incurring unnecessary costs.

Configuring VM Settings

Properly configuring your virtual machine settings is essential to ensure the security, performance, and manageability of your cloud resources. In this section, we will discuss the key aspects of VM configuration across AWS, Azure, and GCP, along with practical examples to guide you through the process.

Network Settings:

Configuring network settings for your VM involves setting up a Virtual Private Cloud (VPC) or Virtual Network, defining subnets, and configuring IP addresses.

- AWS: In Amazon EC2, create a VPC and define subnets within it. Assign an elastic IP address to your VM for a static public IP address, or use the auto-assigned private IP address.

- Azure: In Azure VM, create a Virtual Network and define subnets within it. Assign a static public IP address or use a dynamic private IP address for your VM.

- GCP: In Google Compute Engine, create a VPC network and define subnets within it. Assign an external IP address for a static public IP address, or use the auto-assigned internal IP address.

Security Groups or Firewalls:

Configuring security groups or firewalls is crucial to protect your VM from unauthorized access and to define allowed traffic.

- **AWS:** Create a security group in Amazon EC2, defining inbound and outbound rules to allow specific traffic (e.g., SSH or RDP access) to your VM.

- Example: To allow SSH access to your VM, create a security group with an inbound rule that allows traffic on port 22 from your IP address.

- **Azure:** Create a network security group in Azure, defining inbound and outbound security rules for your VM.

- Example: To allow RDP access to your VM, create a network security group with an inbound rule that allows traffic on port 3389 from your IP address.

- **GCP:** Set up firewall rules in Google Compute Engine, allowing specific traffic to your VM.

- Example: To allow HTTP access to your VM, create a firewall rule that allows traffic on port 80 from any IP address.

Operating System and Disk Size:

Selecting the appropriate operating system and disk size is critical for your VM's performance and compatibility with your applications.

- **AWS:** Amazon EC2 allows you to choose from various Amazon Machine Images (AMIs), which include popular Linux distributions and Windows Server versions. Additionally, you can define the disk size and type (e.g., SSD or HDD) for your VM.

- **Azure:** Azure VM offers a gallery of images, including popular Linux distributions, Windows Server versions, and custom images. You can also define the disk size and type (e.g., premium or standard SSD) for your VM.

- **GCP:** Google Compute Engine provides a selection of public images, including popular Linux distributions and Windows Server versions, as well as custom images. You can also define the disk size and type (e.g., SSD or HDD) for your VM.

Understanding and implementing these VM configuration steps will allow you to ensure your virtual machine is secure, performs optimally, and is compatible with your applications and workloads.

Deploying the Virtual Machine

Deploying a virtual machine is a straightforward process, and each cloud provider offers multiple methods for doing so. In this section, we will explore the deployment process for AWS, Azure, and GCP using their respective consoles, CLI, or SDKs, along with practical examples to help you understand the key steps involved.

AWS EC2:

- Console: To deploy an Amazon EC2 instance, navigate to the AWS Management Console, select the EC2 service, and click "Launch Instance." Follow the wizard to select an Amazon Machine Image (AMI), choose an instance type, configure instance details, storage, and security groups, and finally launch the instance.

- CLI: Using the AWS CLI, you can deploy an EC2 instance with the aws ec2 run-instances command, specifying the required parameters like the AMI ID, instance type, and key pair.

- Example: aws ec2 run-instances --image-id ami-0c55b159cbfafe1f0 --instance-type t2.micro --key-name my-key-pair

Azure VM:

- Console: To deploy an Azure Virtual Machine, navigate to the Azure Portal, click "Create a resource," and select "Virtual Machine." Follow the wizard to configure basic settings, choose an operating system image, select a VM size, and configure networking, storage, and security settings before creating the VM.

- CLI: Using the Azure CLI, you can deploy a VM with the az vm create command, specifying required parameters like the resource group, VM name, OS image, and size.
- Example: az vm create --resource-group myResourceGroup --name myVM --image UbuntuLTS --size Standard_B1s --generate-ssh-keys

GCP Compute Engine:

- Console: To deploy a Google Compute Engine instance, navigate to the Google Cloud Console, select the Compute Engine service, and click "Create Instance." Follow the wizard to configure instance settings, choose an operating system image, select a machine type, and configure networking, storage, and firewall settings before creating the instance.
- CLI: Using the Google Cloud SDK, you can deploy a Compute Engine instance with the gcloud compute instances create command, specifying required parameters like the instance name, zone, machine type, and image.
- Example: gcloud compute instances create my-instance --zone us-central1-a --machine-type n1-standard-1 --image-family debian-9 --image-project debian-cloud

Familiarizing yourself with the deployment process for each cloud provider will allow you confidently launch and manage virtual machines across AWS, Azure, and GCP using their consoles, CLI, or SDKs. This knowledge is essential for effectively utilizing cloud resources and scaling your applications and services as needed.

Accessing and Managing the VM

Once your virtual machine is deployed, you need to know how to access and manage it effectively. In this section, we will discuss accessing your VM using SSH or RDP, and delve into essential management tasks such as monitoring, resizing, and updating your VM instance.

Accessing Your VM:

- SSH (Linux): To access a Linux-based VM, use an SSH client (e.g., OpenSSH, PuTTY) and the VM's public IP address or hostname, along with the appropriate private key for authentication.
- Example: ssh -i my-private-key.pem ubuntu@your-vm-public-ip
- RDP (Windows): To access a Windows-based VM, use a Remote Desktop client (e.g., Microsoft Remote Desktop) and the VM's public IP address or hostname, along with the appropriate username and password or security credentials.
- Example: Open Microsoft Remote Desktop, enter the VM's public IP address, and provide the username and password when prompted.

Monitoring Your VM:

- Monitoring your VM is essential for ensuring optimal performance, resource utilization, and detecting potential issues. Each cloud provider offers monitoring tools and services:
- **AWS:** Use Amazon CloudWatch to monitor metrics such as CPU usage, disk I/O, and network activity for your EC2 instance.
- **Azure:** Use Azure Monitor to view metrics like CPU usage, disk I/O, and network activity for your Azure VM.
- **GCP:** Use Google Cloud Monitoring (formerly Stackdriver) to monitor metrics such as CPU usage, disk I/O, and network activity for your Compute Engine instance.

Resizing Your VM:

- Resizing your VM allows you to adjust its resources (CPU, memory, and storage) to meet changing demands. Each cloud provider offers a method for resizing VM instances:
- AWS: To resize an EC2 instance, stop the instance, modify its instance type, and restart it.
- Azure: To resize an Azure VM, stop the VM, change its size, and start it again.
- GCP: To resize a Compute Engine instance, stop the instance, edit its machine type, and restart it.

Applying Updates: Keeping your VM up-to-date with the latest security patches and software updates is critical for maintaining a secure and stable environment:

- Linux-based VMs: Use the package manager (e.g., apt, yum, zypper) to update your system packages and apply security patches.
- Example: sudo apt update && sudo apt upgrade
- Windows-based VMs: Use Windows Update to check for updates and install them, or configure automatic updates to keep your system up-to-date.

Learning how to access, monitor, resize, and update your virtual machine instances, will allow you maintain a secure, stable, and efficient cloud environment for your applications and services.

Cleaning Up and Terminating the VM

Properly shutting down and deleting your virtual machine when it's no longer needed is crucial for avoiding unnecessary costs and maintaining a clean cloud environment. In this section, we will discuss the importance of VM cleanup and walk through the process of terminating VM instances in AWS, Azure, and GCP.

Importance of VM Cleanup

Virtual machines consume resources and incur costs while they are running, even if they are not actively being used. To optimize resource usage and minimize costs, it's essential to shut down and delete VMs that are no longer needed. Additionally, cleaning up unused resources ensures a more organized and efficient cloud environment.

Terminating VM Instances in AWS, Azure, and GCP

Each cloud provider has a different process for shutting down and deleting VM instances. Here are the steps for each:

AWS EC2:

- Console: Navigate to the AWS Management Console, select the EC2 service, choose the "Instances" section, select the instance you want to terminate, click "Actions," and choose "Instance State" > "Terminate."
- CLI: Use the AWS CLI command aws ec2 terminate-instances and specify the instance ID you want to terminate.
- Example: aws ec2 terminate-instances --instance-ids i-1234567890abcdef0

Azure VM:

- Console: Go to the Azure Portal, navigate to "Virtual Machines," select the VM you want to delete, click on "Overview," and choose "Delete" from the top menu.
- CLI: Use the Azure CLI command az vm delete and specify the resource group and VM name you want to delete.
- Example: az vm delete --resource-group myResourceGroup --name myVM --yes

GCP Compute Engine:

- Console: Navigate to the Google Cloud Console, select the Compute Engine service, choose the "VM instances" section, select the instance you want to delete, click "Delete" at the top of the page, and confirm the deletion.
- CLI: Use the Google Cloud SDK command gcloud compute instances delete and specify the instance name and zone you want to delete.
- Example: gcloud compute instances delete my-instance --zone us-central1-a

Understanding the importance of VM cleanup and learning the process of shutting down and deleting VM instances within AWS, Azure, and GCP, will allow you maintain a cost-efficient and organized cloud environment.

Module 1 Exam Questions

1. Which of the following cloud service providers is the largest in terms of market share?

a) Amazon Web Services (AWS)

b) Microsoft Azure

c) Google Cloud Platform (GCP)

d) IBM Cloud

2. What does the acronym VM stand for in the context of cloud computing?

a) Version Manager

b) Virtual Machine

c) Video Memory

d) Variable Module

3. Which of the following is NOT a type of cloud storage?

a) Object Storage

b) Block Storage

c) File Storage

d) Dynamic Storage

4. Which cloud service provider offers the "S3 Glacier" storage class for long-term data archiving?

a) AWS

b) Azure

c) GCP

d) IBM Cloud

5. What is the primary purpose of Identity and Access Management (IAM) in cloud environments?

a) Data storage optimization

b) Network traffic management

c) User access and permission management

d) Resource tagging

6. Which of the following cloud storage services is offered by Google Cloud Platform (GCP)?

a) Amazon S3

b) Azure Blob Storage

c) Google Cloud Storage

d) IBM Cloud Object Storage

7. In AWS, which service is responsible for providing block storage?

a) Amazon S3

b) Amazon EBS

c) Amazon EFS

d) Amazon Glacier

8. What does the acronym MFA stand for?

a) Multi-Factor Authentication

b) Managed File Access

c) Maximum Firewall Allowance

d) Multiple Firewall Associations

9. Which of the following is a hybrid cloud storage solution?

a) AWS Storage Gateway

b) Google Cloud Storage

c) Amazon S3

d) Azure Blob Storage

10. Which migration method is most suitable when network bandwidth is limited?

a) Online migration

b) Offline migration

c) Hybrid migration

d) None of the above

11. What is the primary function of Azure Managed Disks?

a) Object storage

b) Block storage

c) File storage

d) Archival storage

12. Which of the following cloud providers offers the Compute Engine service?

a) AWS

b) Azure

c) GCP

d) IBM Cloud

13. What is the purpose of tagging resources in a cloud environment?

a) To monitor resource usage

b) To enable multi-factor authentication

c) To organize resources for efficient management

d) To reduce data migration time

14. Which of the following is a cloud-native data migration tool offered by AWS?

a) AWS Database Migration Service

b) Azure Data Migration Service

c) Google Cloud Data Transfer Service

d) None of the above

15. What is the primary purpose of Amazon EFS?

a) Object storage

b) Block storage

c) File storage

d) Archival storage

16. Which of the following is a managed file storage solution offered by Microsoft Azure?

a) Azure Files

b) Azure Blob Storage

c) Azure Managed Disks

d) Azure Archive Storage

17. In cloud environments, which protocol is commonly used for accessing virtual machines remotely on Linux systems?

a) RDP

b) SSH

c) HTTP

d) FTP

18. Which of the following is a factor to consider when choosing a cloud service provider?

a) Market share

b) Features and services

c) Pricing and cost models

d) All of the above

19. What is the primary purpose of setting up billing alerts in a cloud environment?

a) To monitor data migration progress

b) To manage resource tags

c) To track and control cloud spending

d) To improve data security

20. What is the main advantage of using a hybrid cloud storage solution?

a) Reduced storage capacity

b) Integration of on-premises storage with cloud storage

c) Automatic data migration

d) Increased network latency

Answers:

1. b
2. c
3. a
4. d
5. a
6. c
7. b
8. d
9. b
10. a
11. d
12. c
13. a
14. c
15. b
16. a
17. d
18. b
19. c
20. b

Module 2. Cloud Storage Solutions

Introduction to Cloud Storage Solutions

Welcome to Module 2: Cloud Storage Solutions. In this module, we will explore various storage options available in the cloud, their use cases, and best practices for implementation. You will learn about the different storage services offered by major cloud providers such as AWS, Azure, and GCP, and how to select the most appropriate solution for your specific requirements.

Cloud storage solutions provide a flexible, scalable, and cost-effective way to store and manage data. From simple object storage to high-performance block storage and scalable file systems, the cloud offers a wide array of storage options to meet the diverse needs of modern applications and workloads. By understanding the advantages and trade-offs of each storage solution, you can make informed decisions and optimize your cloud environment for performance, cost, and reliability.

Throughout this module, we will provide practical examples and hands-on exercises to help you gain experience in deploying and managing cloud storage services. We will discuss various storage types such as object storage, block storage, and file storage, and delve into the unique features and benefits of each service.

By the end of this module, you will have a solid understanding of the cloud storage landscape and the confidence to choose and implement the right storage solutions for your applications and workloads.

Cloud Storage Types and Use Cases

Object Storage

Object storage is a highly scalable and cost-effective storage solution designed for storing and managing unstructured data, such as images, videos, documents, and backups. In this section, we will explore the fundamentals of object storage services, their benefits, and use cases, with a focus on Amazon S3, Azure Blob Storage, and Google Cloud Storage.

Fundamentals:

Object storage services organize data as objects, which consist of data, metadata, and a unique identifier. This flat structure allows for massive scalability, as objects are distributed across multiple storage nodes and can be accessed directly using their unique identifier. Key features of object storage include:

- **High durability and availability:** By replicating objects across multiple storage nodes or geographic locations, object storage ensures high durability and availability, minimizing the risk of data loss.
- **Scalability:** Object storage is designed to handle virtually unlimited amounts of data, making it an ideal solution for storing large amounts of unstructured data.
- **Cost-effectiveness:** With pay-as-you-go pricing models and features like data tiering and lifecycle policies, object storage services enable cost optimization based on data access patterns and retention requirements.

Benefits:

Object storage services, such as Amazon S3, Azure Blob Storage, and Google Cloud Storage, offer numerous advantages, including:

- **Simple and efficient data management:** Easily store, retrieve, and manage large amounts of unstructured data with a simple RESTful API or web console.
- **Flexible access control:** Secure your data with fine-grained access control policies, user authentication, and encryption.
- **Integration with other cloud services:** Seamlessly integrate object storage with other cloud services for data processing, analytics, and content delivery.

Practical Examples:

- **Amazon S3:** Store images for a web application, and use Amazon CloudFront to deliver the images to users with low latency.
- **Azure Blob Storage:** Store log files generated by an application and use Azure Functions to process the logs for analytics and monitoring purposes.
- **Google Cloud Storage:** Create a backup of database snapshots and configure a lifecycle policy to automatically transition older backups to a lower-cost storage class.

Use Cases:

Object storage services are suitable for a wide range of applications, such as:

- **Content distribution:** Store and distribute static assets like images, videos, and documents for web and mobile applications.
- **Data backups and archiving:** Store backups and archives of databases, virtual machines, and file systems for disaster recovery and compliance purposes.

- **Big data analytics:** Store large datasets for data processing and analytics using cloud-based big data platforms, like Hadoop or Spark.

Understanding the fundamentals, benefits, and use cases of object storage services like Amazon S3, Azure Blob Storage, and Google Cloud Storage, will allow you efficiently store and manage unstructured data in the cloud, while optimizing performance, reliability, and cost.

Block Storage

Block storage services provide low-latency, high-performance storage for use with virtual machines and container-based applications. In this section, we will explore the features, performance characteristics, and applications of popular block storage services, such as Amazon Elastic Block Store (EBS), Azure Managed Disks, and Google Persistent Disks.

Features:

Block storage services offer several key features, including:

- **High performance:** Block storage provides low-latency access to data, making it suitable for applications that require fast read and write operations.
- **Durability and reliability:** By replicating data across multiple storage nodes, block storage services ensure high durability and reliability, reducing the risk of data loss.
- **Snapshots and backups:** Create point-in-time snapshots of your volumes for data protection and disaster recovery purposes.
- **Encryption:** Secure your data at rest and in transit with encryption features provided by the block storage service.

Performance Characteristics:

Block storage services can be optimized for different performance characteristics, such as:

- **IOPS (Input/Output Operations Per Second):** The number of read and write operations per second that a storage volume can handle.
- **Throughput:** The amount of data that can be read or written to a storage volume in a given time period.
- **Latency:** The time it takes to complete a read or write operation to a storage volume.

Selecting the appropriate block storage type and configuration will allow you optimize your storage solution for the specific needs of your applications and workloads.

Practical Examples:

- **Amazon EBS:** Attach an EBS volume to an Amazon EC2 instance to store a database, and use EBS snapshots to create point-in-time backups for disaster recovery.
- **Azure Managed Disks:** Provision a managed disk for an Azure Virtual Machine running a web application, and use Azure Backup to create and manage backups of the disk.
- **Google Persistent Disks:** Attach a persistent disk to a Google Compute Engine instance for storing application data, and use Google Cloud's snapshot feature to create and manage disk snapshots.

Applications:

Block storage services are ideal for a variety of applications, such as:

- **Databases:** Store and manage database files, transaction logs, and indexes with low latency and high performance.

- **File systems:** Create and manage file systems for use with virtual machines or container-based applications.
- **High-performance computing:** Store and process large datasets for scientific research, financial modeling, or media processing.

Understanding the features, performance characteristics, and applications of block storage services like Amazon EBS, Azure Managed Disks, and Google Persistent Disks, will you effectively store and manage data for your applications, while optimizing performance, reliability, and cost.

File Storage

Managed file storage solutions provide shared access to file systems for multiple virtual machines, containers, or applications. In this section, we will learn about popular managed file storage services, including Amazon Elastic File System (EFS), Azure Files, and Google Cloud Filestore, and their use cases in distributed systems and applications.

Features:

Managed file storage solutions offer several key features, such as:

- **Shared access:** Enable multiple clients or instances to access and share data within a single file system.
- **Scalability:** Automatically scale storage capacity as your data grows, without the need for manual provisioning or intervention.
- **Durability and availability:** Ensure high durability and availability through data replication across multiple storage nodes or geographic locations.
- **Access control and security:** Protect your data with access control policies, user authentication, and encryption.

Practical Examples:

- **Amazon EFS:** Create a shared file system for multiple Amazon EC2 instances running a containerized application, and use AWS Identity and Access Management (IAM) to manage access permissions.
- **Azure Files:** Set up a file share accessible by multiple Azure Virtual Machines, and use Azure Active Directory Domain Services (AD DS) to authenticate users and manage access controls.
- **Google Cloud Filestore:** Provision a shared file system for Google Compute Engine instances running a web application, and use Google Cloud Identity and Access Management (IAM) to secure access to the file system.

Use Cases:

Managed file storage services are well-suited for a range of applications and scenarios, such as:

- **Distributed systems:** Provide a shared file system for applications running on multiple virtual machines or containers, enabling efficient data sharing and collaboration.
- **Content management:** Store and manage documents, images, and other assets for content management systems or web applications.
- **Backup and recovery:** Create a centralized repository for backups and archives, allowing for easy data recovery in the event of a disaster or system failure.
- **Software development:** Share source code, build artifacts, and other development resources across development, testing, and production environments.

Understanding the features, practical examples, and use cases of managed file storage solutions like Amazon EFS, Azure Files, and

Google Cloud Filestore, will allow you leverage shared file systems for distributed systems and applications, optimizing performance, reliability, and cost.

Archival Storage

Archival storage solutions provide cost-effective, long-term storage for infrequently accessed data. In this section, we will delve into popular archival storage options, including Amazon S3 Glacier, Azure Archive Storage, and Google Cloud Archive, and their role in data retention and compliance.

Features:

Archival storage solutions offer several key features, such as:

- **Low cost:** Benefit from lower storage costs compared to other storage classes, making archival storage suitable for large volumes of infrequently accessed data.
- **Durability:** Ensure high durability and data protection through redundancy and replication across multiple storage nodes or geographic locations.
- **Data retrieval:** Access your archived data with configurable retrieval times, ranging from minutes to hours, depending on your needs.
- **Compliance and security:** Meet regulatory requirements and secure your data with features like data encryption, access control policies, and audit logging.

Practical Examples:

- **Amazon S3 Glacier:** Store historical log files, backups, and other infrequently accessed data in Amazon S3 Glacier, and use Amazon S3 Select to query and retrieve specific data when needed.
- **Azure Archive Storage:** Migrate long-term backups and archives to Azure Archive Storage, and use Azure Backup

or Azure Data Factory to automate the backup and archiving process.

- **Google Cloud Archive:** Archive large datasets, such as medical records or financial data, in Google Cloud Archive, and use Google Cloud Storage Transfer Service to move data between storage classes.

Use Cases:

Archival storage services are ideal for a variety of data retention and compliance scenarios, such as:

- **Long-term backups:** Store backups of critical systems and data for extended periods to meet data retention requirements and ensure disaster recovery.
- **Compliance and regulatory archiving:** Archive data subject to legal, financial, or industry-specific regulations, ensuring compliance with data retention and security requirements.
- **Historical data analysis:** Preserve historical data for trend analysis, research, or business intelligence purposes, enabling insights into past performance or events.
- **Digital preservation:** Store and protect digital assets, such as media files, documents, or scientific data, for long-term access and preservation.

Understanding the features, practical examples, and use cases of cost-effective, long-term storage options like Amazon S3 Glacier, Azure Archive Storage, and Google Cloud Archive, will allow you effectively retain and manage your organization's data, while meeting compliance requirements and optimizing storage costs.

Hybrid and Multi-Cloud Storage

Hybrid and multi-cloud storage solutions enable seamless integration of on-premises storage with cloud storage, or the use of multiple cloud providers for storage. In this section, we will examine strategies and solutions for hybrid and multi-cloud storage, including AWS Storage Gateway, Azure StorSimple, and Google Cloud Anthos.

Features: Hybrid and multi-cloud storage solutions offer several key features, such as:

- **Storage integration:** Connect your on-premises storage systems with cloud storage services, allowing for seamless data movement and synchronization.
- **Multi-cloud support:** Leverage the strengths of different cloud providers, distributing your storage workloads across multiple platforms for increased flexibility, resilience, and cost optimization.
- **Data management:** Simplify data management with centralized control, policy-based data movement, and automated tiering of data between storage classes or providers.
- **Security and compliance:** Secure your data both on-premises and in the cloud with features like data encryption, access control policies, and audit logging.

Practical Examples:

- **AWS Storage Gateway:** Deploy a virtual or physical gateway appliance on-premises to connect your existing storage systems with Amazon S3, Amazon EBS, or Amazon Glacier, enabling seamless data movement and backup.
- **Azure StorSimple:** Implement a StorSimple device or virtual appliance on-premises to connect your local storage with Azure Blob Storage, and use features like tiering, deduplication, and compression to optimize storage utilization and costs.

- **Google Cloud Anthos:** Deploy Anthos to manage your multi-cloud storage environment, allowing for consistent data and policy management across on-premises storage and multiple cloud providers, such as Google Cloud Storage, Amazon S3, and Azure Blob Storage.

Use Cases:

Hybrid and multi-cloud storage solutions are well-suited for a variety of scenarios, such as:

- **Data migration:** Facilitate the migration of data from on-premises storage to the cloud, or between cloud providers, with minimal downtime and disruption.
- **Backup and disaster recovery:** Implement a hybrid or multi-cloud backup and disaster recovery strategy, storing copies of your data both on-premises and in the cloud for added redundancy and protection.
- **Data sharing and collaboration:** Enable seamless access to data across your organization, regardless of whether it is stored on-premises or in the cloud, promoting collaboration and productivity.
- **Cost optimization:** Leverage the cost benefits of different storage classes or cloud providers, optimizing storage costs based on data access patterns, performance requirements, and other factors.

Examining the strategies and solutions for integrating on-premises storage with cloud storage or leveraging multiple cloud providers for storage, including AWS Storage Gateway, Azure StorSimple, and Google Cloud Anthos, will allow you optimize your storage infrastructure, enhance data management, and ensure data security and compliance across your hybrid or multi-cloud environment.

Data Migration Strategies

Migrating data to the cloud is a critical component of any cloud transformation journey. A well-planned and executed data migration strategy ensures the seamless transfer of data from on-premises storage or other cloud providers to your chosen cloud platform while minimizing downtime, preserving data integrity, and maintaining security and compliance. In this section, we will delve into various data migration strategies, providing detailed information and practical examples that will help you select and implement the most effective approach for your organization.

Key factors to consider when planning a data migration include data size, access patterns, dependencies, and migration scope. By understanding the intricacies of different migration methods, such as online migration, offline migration, or hybrid migration, you can choose the best-fit approach based on your organization's specific requirements, such as downtime tolerance, network bandwidth, and data consistency needs.

We will also explore cloud-native migration tools provided by leading cloud service providers, such as AWS Database Migration Service, Azure Data Migration Service, and Google Cloud Data Transfer Service, which simplify and streamline the migration process. Additionally, we will discuss the importance of implementing appropriate security measures during data migration, including data encryption, access control, and monitoring.

Lastly, this section will emphasize the significance of validating and optimizing the migration process, ensuring data integrity and consistency, and continuously improving the migration process through performance monitoring, tuning, and adjustments.

Assessing and Planning Data Migration

Before embarking on any data migration project, it is crucial to conduct a thorough assessment and create a comprehensive plan to ensure a smooth migration process. This involves evaluating the data to be migrated, determining the migration scope, and considering factors such as data size, access patterns, and dependencies.

Detailed Information:

- **Data Evaluation:** Start by cataloging all data sources and types within your organization. This may include databases, file systems, applications, and data warehouses. Identify the data formats, such as structured, unstructured, or semi-structured data, and understand the data's sensitivity, as this will impact security measures during migration.

- **Migration Scope:** Determine which data sets need to be migrated and prioritize them based on factors such as business importance, data dependencies, and the complexity of migration. Consider whether a full migration, incremental migration, or a phased migration is the best approach for your organization.

- **Data Size and Access Patterns:** Estimate the size of the data to be migrated and analyze access patterns to understand the performance requirements and potential impact on network bandwidth during migration. This will help in selecting the appropriate migration method and optimizing the process to minimize downtime.

  Practical Example:

  - A medium-sized e-commerce company plans to migrate its on-premises customer database to a cloud-based database service. The company's IT team starts by cataloging all data sources, including customer information, order history, and product inventory. They determine that the customer database and order history need to be migrated as a priority, while the product inventory data can be migrated in a later phase.

- The team estimates the total size of the customer database and order history to be 2 TB, with frequent read and write operations. Based on this analysis, they decide to perform an incremental migration to minimize downtime and reduce the impact on network bandwidth. The migration plan includes detailed steps, such as data backup, migration method selection, security measures, and validation procedures to ensure a successful migration process.

- **Data Dependencies:** Identify any dependencies between the data sets to be migrated and other systems or applications within your organization. Understanding these dependencies is crucial to ensure seamless integration and avoid disruptions during migration. Consider the need for data transformation or reformatting to maintain compatibility with the target cloud storage or database services.

- **Create a Detailed Migration Plan:** Based on the assessment, develop a comprehensive migration plan that outlines the steps, timelines, and resources required for a successful migration. Include pre-migration tasks such as data backup, data transformation, and security configurations, as well as post-migration tasks like data validation, performance monitoring, and optimization. Assign roles and responsibilities to team members and establish clear communication channels to ensure smooth execution.

  Practical Example:

  - Continuing with the e-commerce company example, the IT team identifies dependencies between the customer database and the company's marketing automation system. They realize that data transformation is necessary to maintain compatibility with the target cloud database service.

- The team creates a detailed migration plan that includes pre-migration tasks like backing up the data, transforming it to the required format, and setting up necessary security configurations. They also plan post-migration tasks such as data validation, performance monitoring, and optimization. To ensure smooth execution, they assign roles and responsibilities to team members and establish communication channels for reporting progress and addressing any issues that may arise during the migration process.

Choosing the Right Migration Method

Selecting the appropriate migration method is crucial to the success of any data migration project. Various methods exist, such as online migration, offline migration, or hybrid migration. The choice depends on factors like downtime tolerance, network bandwidth, and data consistency requirements. Understanding the nuances of each method will help you make an informed decision that meets your organization's specific needs.

Detailed Information:

- **Online Migration:** Online migration, also known as live migration, involves transferring data to the cloud while the source systems continue to operate. This method minimizes downtime and is suitable for organizations that require continuous access to their data during migration. However, it may consume significant network bandwidth and might not be suitable for organizations with limited connectivity or strict data consistency requirements.

  Practical Example: A financial services firm wants to migrate its transaction database to the cloud but cannot afford any downtime during the migration process. They choose online migration, which allows them to continue processing transactions while the data is being transferred to the cloud.

- **Offline Migration:** Offline migration involves transferring data to the cloud while the source systems are temporarily offline or unavailable. This method can be faster and more straightforward, as it avoids potential conflicts or inconsistencies in the data. However, it requires a period of downtime, which may not be acceptable for organizations with strict availability requirements.

  Practical Example: A manufacturing company wants to migrate its production data to the cloud during a scheduled maintenance window. They choose offline migration, as they can afford downtime during the maintenance period and want to ensure data consistency.

- **Hybrid Migration:** Hybrid migration is a combination of online and offline migration methods. It typically involves an initial offline migration, followed by incremental online migrations to synchronize any changes made to the data during the migration process. This approach balances the need for minimal downtime and data consistency, but may require more planning and coordination.

  Practical Example: A retail company wants to migrate its inventory management system to the cloud with minimal downtime and data consistency. They perform an initial offline migration during a low-traffic period, then use online migration to synchronize any changes made to the inventory data during the migration process.

When choosing the right migration method, consider your organization's specific requirements, such as downtime tolerance, network bandwidth limitations, and data consistency needs.

Assess the trade-offs and select the approach that best aligns with your objectives and constraints.

Leveraging Cloud-native Migration Tools

Utilizing cloud-native migration tools provided by AWS, Azure, and GCP can significantly simplify the migration process, improve efficiency, and reduce the potential for errors. These tools are designed to work seamlessly with their respective cloud platforms, facilitating a smooth migration experience while ensuring optimal performance and compatibility.

Detailed Information:

- **AWS Database Migration Service (DMS):** AWS DMS is a managed service that helps you migrate your databases to AWS with minimal downtime. It supports homogeneous migrations (e.g., Oracle to Oracle) and heterogeneous migrations (e.g., Oracle to Amazon Aurora). AWS DMS also supports continuous data replication for ongoing synchronization between source and target databases.

  Practical Example: An e-commerce company wants to migrate its Oracle database to Amazon Aurora. They use AWS DMS to perform a heterogeneous migration, minimizing downtime and ensuring that their application remains operational during the migration process.

- **Azure Data Migration Service (DMS):** Azure DMS is a fully managed, comprehensive solution for migrating various types of data to Azure cloud services. It supports migration from multiple source types, including SQL Server, MySQL, PostgreSQL, and others, to Azure SQL Database, Azure SQL Managed Instance, and Azure Database for MySQL, PostgreSQL, and MariaDB.

  Practical Example: A software development company wants to migrate its on-premises SQL Server database to Azure SQL Database. They use Azure DMS to perform the migration, taking advantage of its built-in performance monitoring and compatibility checks to ensure a smooth transition to the cloud.

- **Google Cloud Data Transfer Service:** Google Cloud Data Transfer Service helps you move your data from various sources, such as AWS S3, HTTP/HTTPS, or on-premises storage systems, to Google Cloud Storage. It supports online and offline data transfer methods, as well as various data formats, including CSV, JSON, and Avro.

  Practical Example: A media company wants to migrate its large multimedia files from an on-premises storage system to Google Cloud Storage. They use Google Cloud Data Transfer Service to perform an efficient and secure data transfer, leveraging its support for parallel uploads and automatic retries in case of failures.

Leveraging these cloud-native migration tools will allow you streamline the data migration process and minimize the risks associated with manual migrations. These tools not only help you save time and resources but also ensure compatibility with your target cloud platform, enabling you to focus on optimizing your cloud environment for maximum value.

Ensuring Data Security and Compliance

Data security and compliance are paramount during the migration process. Implementing appropriate security measures, such as data encryption, access control, and monitoring, helps protect sensitive information and ensures that you maintain compliance with relevant regulations throughout the migration.

Detailed Information:

- **Data Encryption:** Encrypting data during transit and at rest ensures that sensitive information remains protected even if it's intercepted or accessed by unauthorized individuals. Cloud service providers offer built-in encryption mechanisms, such as AWS Key Management Service (KMS), Azure Key Vault, or Google Cloud Key Management Service (KMS), to facilitate secure encryption key management and seamless integration with their respective storage services.

Practical Example: A healthcare organization needs to migrate patient records to the cloud while ensuring that the data remains encrypted. They use AWS KMS to manage encryption keys and configure their Amazon S3 buckets to automatically encrypt all data at rest using the provided keys.

- **Access Control:** Implementing strict access control measures during the migration process helps prevent unauthorized access to your data. This can be achieved by leveraging Identity and Access Management (IAM) features, roles, and policies provided by AWS, Azure, or GCP to restrict access to specific resources and actions based on the principle of least privilege.

  Practical Example: A financial institution wants to migrate their customer data to Google Cloud Storage. They use Google Cloud IAM to create a custom role that grants access to only the necessary resources and actions required for the migration process, ensuring that only authorized personnel can perform the migration.

- **Monitoring:** Continuously monitoring your data migration process helps identify potential security threats and maintain compliance with relevant regulations. Cloud providers offer monitoring and logging services, such as AWS CloudTrail, Azure Monitor, or Google Cloud Logging, to track activity within your cloud environment and alert you to any suspicious activities.

  Practical Example: A retail company is migrating sensitive customer data to Azure Blob Storage. They use Azure

Monitor to set up alerts for any unusual access patterns or unauthorized attempts to access the data, enabling them to respond quickly to potential security threats.

Implementing these security measures during data migration will allow you to ensure the protection of sensitive information and maintain compliance with industry regulations. This not only helps to maintain customer trust but also mitigates the risk of costly fines and reputational damage due to data breaches or non-compliance.

Validating and Optimizing the Migration Process

Performing thorough testing and validation of the migrated data is essential to ensure data integrity and consistency. Continuously optimizing the migration process through performance monitoring, tuning, and adjustments helps achieve a smooth and efficient migration experience.

Detailed Information:

- **Data Validation:** After migrating your data to the cloud, it's crucial to validate the accuracy and consistency of the data. This can be done through techniques such as data sampling, checksum comparisons, or running integrity checks using tools provided by your cloud service provider. By validating the migrated data, you ensure that no data corruption or loss occurred during the migration process.

  Practical Example: A company has migrated their product database from an on-premises server to Amazon RDS. To validate the data migration, they run a series of SQL queries and compare the results with the original data to ensure data consistency and integrity.

- **Performance Monitoring:** Continuously monitoring the performance of the migration process allows you to identify bottlenecks and areas for improvement. Cloud providers offer tools and services for performance monitoring, such as AWS CloudWatch, Azure Monitor, or Google Cloud

Monitoring, which can be used to track key performance metrics and identify potential issues.

Practical Example: An organization migrating a large dataset to Google Cloud Storage notices that the migration process is taking longer than expected. By analyzing the performance metrics in Google Cloud Monitoring, they identify a network bottleneck as the cause and take appropriate measures to address the issue.

- **Tuning and Adjustments:** As you monitor the migration process, it's essential to make adjustments and optimizations based on the insights gained from performance monitoring. This may include changing the migration method, adjusting parallelism, modifying the data transfer rate, or fine-tuning the configuration settings of the migration tools.

  Practical Example: A company migrating their data to Azure Blob Storage realizes that the data transfer rate is lower than anticipated. To optimize the migration process, they adjust the number of parallel data transfer threads and increase the bandwidth allocated to the migration, resulting in a significant improvement in the overall migration speed.

Validating and continuously optimizing the migration process will allow you to ensure a successful and efficient migration experience, minimizing downtime and reducing the risk of data corruption or loss. This also allows you to fine-tune the migration process to meet your specific needs and requirements, ultimately ensuring a smooth transition to the cloud.

Practical: Configuring and Using Cloud Storage Services

In this section we will delve into the practical aspects of working with cloud storage solutions. As businesses of all sizes increasingly rely on cloud-based infrastructure, it is essential for IT professionals to gain hands-on experience in managing cloud storage services effectively. This section aims to provide you with the necessary knowledge and skills to configure, use, and manage cloud storage services across various platforms such as Amazon Web Services (AWS) S3, Google Cloud Storage, and Microsoft Azure Blob Storage.

Throughout this section, we will explore real-world scenarios that demonstrate the benefits of cloud storage services and the challenges that may arise when implementing and managing them. You will learn how to navigate the cloud storage ecosystem, from setting up accounts and uploading data to integrating these services into your applications and ensuring the security of your stored data.

We will also emphasize the importance of monitoring and optimizing cloud storage usage to maintain cost-efficiency and ensure that your cloud storage solution remains scalable, reliable, and secure. By the end of this section, you will be equipped with a solid understanding of cloud storage services and practical experience that can be directly applied in real-world situations.

Let's dive into the hands-on exercises and practical examples that will empower you to become proficient in configuring and using cloud storage services across various platforms.

Setting up cloud storage accounts

Setting up cloud storage accounts: In this practical, you will gain hands-on experience in creating and configuring cloud storage accounts across various platforms, such as Amazon Web Services (AWS) S3, Google Cloud Storage, and Microsoft Azure Blob Storage. The process of setting up an account on each platform may slightly differ, but the key concepts remain consistent. Let's look at each platform in more detail:

- **Amazon Web Services (AWS) S3:** Amazon Simple Storage Service (S3) is a popular object storage service designed for storing and retrieving data at any scale. To set up an AWS S3 account, you will first need to sign up for an AWS account. Once signed up, you can access the AWS Management Console, navigate to the S3 service, and create a new S3 bucket. Buckets are the fundamental containers for data storage in AWS S3. You will need to choose a unique bucket name, select a region, and configure bucket properties like versioning and access control.

  Practical example: You create a new S3 bucket named "my-project-files" in the US East (N. Virginia) region to store project-related data.

- **Google Cloud Storage:** This object storage service is part of the Google Cloud Platform (GCP) and is designed for storing and serving large amounts of unstructured data. You will need to sign up for a GCP account and access the Google Cloud Console. From there, you can create a new storage bucket in the Cloud Storage section, choose a globally unique bucket name, select a storage class, and define access control settings.

  Practical example: You create a new storage bucket named "my-thesis-data" in the Multi-Regional storage class, with the location set to the United States, to store research data for your thesis.

- **Microsoft Azure Blob Storage:** Azure Blob Storage is an object storage service provided by Microsoft Azure, designed for storing large amounts of unstructured data. You will need to sign up for an Azure account and access the Azure Portal. Within the portal, you can create a new storage account and select the "Blob storage" account kind. You will need to choose a unique account name, select a subscription, create or select a resource group, and choose a location for the storage account. Once the storage account is created, you can navigate to the "Containers" section and create a new container to store your data.

  Practical example: You create a new storage account named "myappdata" in the East US region, then creates a container named "user-uploads" to store user-uploaded files for your web application.

Completing this practical exercise will allow you to become familiar with the process of setting up cloud storage accounts on various platforms and will be able to confidently manage your cloud storage resources.

Uploading and downloading data

Uploading and downloading data: In this practical exercise, you will learn various methods to upload and download data to and from cloud storage services. We will explore the use of web interfaces, command-line tools, and APIs for managing data on platforms like Amazon Web Services (AWS) S3, Google Cloud Storage, and Microsoft Azure Blob Storage. Additionally, this exercise will cover managing permissions and access controls for data stored in the cloud.

Let's dive into each method and provide practical examples:

- **Web interfaces:** Cloud storage platforms provide user-friendly web interfaces that allow users to easily upload and download files. You will learn how to navigate through the web interface of each platform and perform data management tasks.

- **AWS S3:** You can use the AWS Management Console to upload files to your S3 bucket by clicking on the "Upload" button, selecting files, and configuring access permissions. Similarly, you can download files by selecting the desired file and clicking on the "Download" button.
- **Google Cloud Storage:** You can upload files to your storage bucket using the Google Cloud Console by clicking on the "Upload files" button and selecting files from your local system. To download files, you can simply click on the file name and select "Download."
- **Azure Blob Storage:** In the Azure Portal, you can upload files to your containers by clicking on the "Upload" button, choosing files from your local system, and configuring access permissions. Downloading files can be achieved by selecting the desired file and clicking on the "Download" button.
- **Command-line tools:** You will learn how to use command-line tools provided by each cloud platform to interact with your cloud storage resources.
- **AWS S3:** You will learn how to use the AWS Command Line Interface (CLI) to upload files to your S3 bucket using the "aws s3 cp" command and download files using the "aws s3 sync" command.
- **Google Cloud Storage:** The Google Cloud SDK provides the "gsutil" command-line tool, which you can use to upload files using the "gsutil cp" command and download files with the "gsutil rsync" command.
- **Azure Blob Storage:** You will use the Azure CLI to upload files to your containers using the "az storage blob upload"

command and download files with the "az storage blob download" command.

- **APIs:** Cloud storage providers offer APIs to integrate their services into applications. You will learn how to use SDKs and APIs for each platform to upload and download files programmatically.

- **AWS S3:** You will explore the AWS SDK for various programming languages (e.g., Python, Java, Node.js) and learn how to use the "upload" and "download" methods for managing files in your S3 bucket.

- **Google Cloud Storage:** You will use the Google Cloud Storage Client Libraries for different languages (e.g., Python, Java, Node.js) to learn how to upload and download files programmatically.

- **Azure Blob Storage:** The Azure Storage SDKs for various languages (e.g., Python, Java, .NET) will help you understand how to upload and download files using the "upload_blob" and "download_blob" methods.

Completing this practical exercise will allow you to be well-versed in uploading and downloading data using different methods and managing access controls for your cloud storage resources across various platforms.

Implementing versioning and lifecycle policies

Implementing versioning and lifecycle policies: In this practical exercise, you will learn how to enable versioning and set up lifecycle policies for your cloud storage. These features help maintain data integrity, facilitate data recovery, and optimize storage costs by automatically managing older versions of files and expiring them when necessary. We will explore versioning and lifecycle policies on platforms like Amazon Web Services (AWS) S3, Google Cloud Storage, and Microsoft Azure Blob Storage. Let's dive into each platform and provide practical examples:

Amazon Web Services (AWS) S3:

- **Versioning:** You will learn how to enable versioning for your S3 buckets using the AWS Management Console. This can be done by navigating to the bucket's properties and enabling the "Versioning" feature. Once enabled, AWS S3 will automatically store all versions of an object, including all writes and deletes, in the bucket.
  Practical example: You enable versioning for your "my-project-files" S3 bucket. When you accidentally overwrite an important file, you can easily recover the previous version through the AWS Management Console.

- **Lifecycle policies:** You will be guided through the process of creating and configuring lifecycle policies in the AWS Management Console. Lifecycle policies allow users to define actions that S3 should take on objects, such as transitioning objects to other storage classes or deleting them after a specific period.
  Practical example: You set up a lifecycle policy for your "my-project-files" S3 bucket that transitions files older than 30 days to the cheaper "S3 One Zone-Infrequent Access" storage class and deletes them after 365 days.

Google Cloud Storage:

- **Versioning:** You will learn how to enable object versioning for your Google Cloud Storage buckets using the Google Cloud Console. You can do this by navigating to the bucket's details page and enabling the "Object versioning" feature. Once enabled, Google Cloud Storage will maintain all versions of an object, allowing for easy recovery of previous versions.
  Practical example: You enable object versioning for your "my-thesis-data" bucket. When a crucial file is mistakenly deleted, you can recover it by accessing previous versions through the Google Cloud Console.

- **Lifecycle policies:** You will learn how to create and configure lifecycle policies for your Google Cloud Storage buckets using the Google Cloud Console. These policies allow users to define conditions and actions for managing objects, such as transitioning objects to other storage classes or deleting them after a specific period.
  Practical example: You set up a lifecycle policy for your "my-thesis-data" bucket that transitions files older than 60 days to the "Nearline" storage class and deletes them after 180 days.

Microsoft Azure Blob Storage:

- **Versioning:** You will learn how to enable versioning for your Azure Blob Storage containers by configuring the "Blob versioning" feature at the storage account level. This can be done using the Azure Portal by navigating to the storage account's settings and enabling the "Blob versioning" option. Once enabled, Azure Blob Storage will store all versions of a blob, allowing for easy recovery of previous versions.
  Practical example: You will enable versioning for you "user-uploads" container. When a user accidentally overwrites a critical file, you can retrieve the previous version through the Azure Portal.

- **Lifecycle policies:** You will learn how to create and configure lifecycle management policies for your Azure Blob Storage containers using the Azure Portal. Lifecycle management policies allow users to define rules that automate the transition of blobs to other access tiers or delete them after a specific period.
  Practical example: You set up a lifecycle management policy for your "user-uploads" container that moves blobs to the "Cool" access tier after 30 days of inactivity and deletes them after 90 days.

Completing this practical exercise will allow you gain a solid understanding of versioning and lifecycle policies across various cloud storage platforms. You will learn how to enable versioning for your storage resources, ensuring that you can recover previous versions of files in case of accidental deletion or overwrite. Additionally, you will learn how to create and configure lifecycle policies to automatically manage older versions of files and expire them as needed, optimizing storage costs and maintaining a clean and efficient storage environment.

Integrating cloud storage with applications

Integrating cloud storage with applications: In this practical exercise, you will learn how to connect cloud storage services to your applications for seamless data access, storage, and retrieval. This will involve using SDKs and APIs provided by cloud storage providers, such as Amazon Web Services (AWS) S3, Google Cloud Storage, and Microsoft Azure Blob Storage. Moreover, you will gain an understanding of best practices for data access and performance. Let's explore each platform and provide practical examples:

Amazon Web Services (AWS) S3:

- **SDKs and APIs:** You will learn how to use the AWS SDKs for various programming languages (e.g., Python, Java, Node.js) to integrate AWS S3 with your applications. You will be guided through the process of authenticating with the AWS S3 service, uploading, and downloading files programmatically, and managing access controls.
  Practical example: You will develop a Python-based web application that allows users to upload images. You use the AWS SDK for Python (Boto3) to securely store these images in your "user-images" S3 bucket and retrieve them for display in the application.

- **Best practices:** You will learn the best practices for data access and performance, such as caching content using

AWS CloudFront, using S3 transfer acceleration for faster uploads, and optimizing object keys for better performance.

Google Cloud Storage:

- **SDKs and APIs:** You will learn how to use Google Cloud Storage Client Libraries for different languages (e.g., Python, Java, Node.js) to integrate Google Cloud Storage with your applications. You will be guided through the process of authenticating with the service, uploading and downloading files programmatically, and managing access controls.
  Practical example: You create a Node.js-based application that processes and stores large datasets. You use the Google Cloud Storage Client Library for Node.js to securely store the processed data in your "processed-data" bucket and retrieve it when needed for further analysis.

- **Best practices:** You will learn best practices for data access and performance, such as using signed URLs for temporary access to objects, caching content using Google Cloud CDN, and using parallel composite uploads for faster uploads.

Microsoft Azure Blob Storage:

- **SDKs and APIs:** You will learn how to use the Azure Storage SDKs for various languages (e.g., Python, Java, .NET) to integrate Azure Blob Storage with your applications. You will be guided through the process of authenticating with the Azure Blob Storage service, uploading and downloading files programmatically, and managing access controls.
  Practical example: You develop a Java-based mobile application that allows users to store and share documents. You use the Azure Storage SDK for Java to securely store these documents in your "user-documents" container and retrieve them for sharing within the app.

- **Best practices:** You will learn best practices for data access and performance, such as using Azure CDN for caching content, optimizing access tiers for cost and performance, and using the Azure Import/Export service for transferring large amounts of data.

Completing this practical exercise will allow you to be well-equipped to integrate cloud storage services with your applications for seamless data access, storage, and retrieval. You will gain hands-on experience using SDKs and APIs provided by cloud storage providers and understand the best practices for data access and performance.

Monitoring and managing cloud storage

Monitoring and managing cloud storage: In this final part of the practical, you will be introduced to monitoring tools and techniques for cloud storage services on platforms like Amazon Web Services (AWS) S3, Google Cloud Storage, and Microsoft Azure Blob Storage. You will learn how to track usage, set up alerts, and troubleshoot any issues that may arise with your cloud storage solutions. Let's explore each platform and provide practical examples:

Amazon Web Services (AWS) S3:

- **Monitoring tools and techniques:** You will learn how to use Amazon CloudWatch, a monitoring service for AWS resources, to track usage and performance metrics for your S3 buckets. You will be guided through the process of creating CloudWatch dashboards, monitoring specific metrics, and setting up alarms for predefined thresholds.
  Practical example: You set up a CloudWatch dashboard to monitor the total number of objects, request rates, and data transfer rates for your "user-images" S3 bucket. You configure an alarm to notify you when the request rate exceeds a predefined threshold, helping you identify potential performance bottlenecks.

- **Troubleshooting:** You will learn how to use AWS S3 access logs, object-level logging, and AWS Trusted Advisor to identify and troubleshoot any issues that may arise with your cloud storage solutions.

Google Cloud Storage:

- **Monitoring tools and techniques:** You will learn how to use Google Cloud Monitoring (formerly Stackdriver) to monitor your Google Cloud Storage buckets. You will be guided through the process of creating custom dashboards, tracking specific metrics, and setting up alert policies based on predefined conditions.
  Practical example: You create a custom dashboard in Google Cloud Monitoring to track the total size, request count, and error rates for your "processed-data" bucket. You configure an alert policy to notify you when the error rate exceeds a specific threshold, helping you identify potential issues.

- **Troubleshooting:** You will learn how to use Google Cloud Storage logs, audit logs, and the Google Cloud Operations suite (formerly Stackdriver) to identify and troubleshoot any issues that may arise with your cloud storage solutions.

Microsoft Azure Blob Storage:

- **Monitoring tools and techniques:** You will learn how to use Azure Monitor, a comprehensive monitoring service for Azure resources, to track usage and performance metrics for your Azure Blob Storage containers. They will be guided through the process of creating custom workbooks, monitoring specific metrics, and setting up alert rules based on predefined conditions.
  Practical example: You create a custom workbook in Azure Monitor to track the total capacity, ingress, egress, and error rates for your "user-documents" container. You configure an alert rule to notify you when the egress rate

exceeds a specific threshold, helping you identify potential bandwidth bottlenecks.

- **Troubleshooting:** You will learn how to use Azure Blob Storage logs, Azure Storage Analytics, and Azure Advisor to identify and troubleshoot any issues that may arise with your cloud storage solutions.

Completing this practical exercise will allow you gain a solid understanding of monitoring and managing cloud storage services across various platforms. You will learn how to track usage and performance metrics, set up alerts to proactively address potential issues, and utilize various tools and techniques to troubleshoot and resolve problems with your cloud storage solutions.

Module 2 Exam Questions

1. Which of the following cloud storage services is provided by Amazon Web Services (AWS)?

a. Google Cloud Storage

b. Azure Blob Storage

c. S3

d. None of the above

2. Which feature of cloud storage allows for easy recovery of previous versions of files?

a. Versioning

b. Lifecycle policies

c. Caching

d. Access control

3. In AWS S3, what is the purpose of a lifecycle policy?

a. To enable versioning

b. To manage older versions of files and expire them as needed

c. To store all versions of an object

d. To optimize object keys for better performance

4. When working with Google Cloud Storage, which of the following tools can be used for monitoring?

a. AWS CloudWatch

b. Azure Monitor

c. Google Cloud Monitoring

d. None of the above

5. Which language is NOT supported by AWS SDKs for integrating AWS S3 with applications?

a. Python

b. Java

c. Ruby

d. Swift

6. Which cloud platform uses "containers" for storage?

a. Amazon Web Services

b. Google Cloud Storage

c. Microsoft Azure Blob Storage

d. None of the above

7. What is a practical use case of versioning in cloud storage?

a. Recovering a mistakenly deleted file

b. Transitioning objects to other storage classes

c. Reducing request rates

d. Caching content

8. Which feature in cloud storage can help reduce costs by transitioning objects to cheaper storage classes?

a. Versioning

b. Access control

c. Caching

d. Lifecycle policies

9. What tool can be used for monitoring Azure Blob Storage?

a. AWS CloudWatch

b. Azure Monitor

c. Google Cloud Monitoring

d. None of the above

10. In the context of cloud storage, what does CDN stand for?

a. Cloud Data Node

b. Content Delivery Network

c. Cloud Deployment Network

d. Container Data Network

11. Which of the following is a cloud storage service provided by Google?

a. AWS S3

b. Google Cloud Storage

c. Azure Blob Storage

d. None of the above

12. When using APIs to access cloud storage services, which of the following is typically required for authentication?

a. API key

b. Username

c. Password

d. Email address

13. What is the purpose of Azure Import/Export service?

a. Monitoring Azure Blob Storage

b. Implementing versioning

c. Transferring large amounts of data

d. Transitioning objects to other storage classes

14. In which cloud storage platform can you create "buckets" for storage?

a. AWS S3

b. Google Cloud Storage

c. Azure Blob Storage

d. Both a and b

15. Which cloud platform uses "blobs" for storage?

a. Amazon Web Services

b. Google Cloud Storage

c. Microsoft Azure Blob Storage

d. None of the above

16. Which of the following is a benefit of using cloud storage?

a. Scalability

b. Cost-effectiveness

c. Data redundancy

d. All of the above

17. What is the main purpose of using signed URLs in cloud storage?

a. To enable versioning

b. To provide temporary access to objects

c. To transition objects to other storage classes

d. To cache content

18. Which of the following is a commonly used cloud storage monitoring technique?

a. Using versioning

b. Creating custom dashboards

c. Implementing access control

d. Setting up a CDN

19. Which of the following cloud storage services is provided by Microsoft?

a. AWS S3

b. Google Cloud Storage

c. Azure Blob Storage

d. None of the above

20. What is the purpose of using a CDN in the context of cloud storage?

a. To enable versioning

b. To provide temporary access to objects

c. To cache content and improve delivery performance

d. To transition objects to other storage classes

Answers:

1. c
2. a
3. b
4. c
5. d
6. c
7. a
8. d
9. b
10. b
11. b
12. a
13. c
14. d
15. c
16. d
17. b
18. b
19. c
20. c

Module 3. Cloud Networking

Introduction to Cloud Networking

Welcome to Module 3. In this module, we will introduce you to the essential concepts and principles of cloud networking, which form the backbone of any cloud-based infrastructure. As organizations increasingly migrate their operations and applications to the cloud, understanding the fundamentals of cloud networking becomes crucial for IT professionals and developers alike.

Cloud networking refers to the interconnections between various components and resources within a cloud environment, enabling communication and data exchange between these entities. As part of a cloud-based infrastructure, networking ensures the seamless integration of different services, applications, and data storage options. This module will help you gain a solid understanding of the underlying technologies and best practices required to design, implement, and manage efficient and secure cloud networks.

We will begin with an overview of the key components and architecture of cloud networks, including virtual networks, subnets, and gateways. This will provide you with a foundational understanding of how cloud networks are structured and how they enable communication between different resources within the cloud environment.

Next, we will explore practical examples of cloud networking in action. We will look at various use cases, such as deploying multi-tier applications, connecting on-premises data centers to cloud services, and implementing secure communication channels using virtual private networks (VPNs). These examples will help you understand the real-world applications of cloud networking concepts and will serve as a basis for the hands-on exercises in the module.

By the end of this module, you will have a strong grasp of cloud networking fundamentals, enabling you to design and implement effective cloud-based solutions for your organization. This knowledge will not only help you create more efficient and secure cloud infrastructures but also contribute to the overall success of your organization's cloud migration and adoption efforts.

So, let's dive into the fascinating world of cloud networking and learn how to harness its power to build and optimize cloud-based infrastructures!

Cloud Network Components and Architecture

In this section, we will delve into the building blocks and key principles that shape the design and implementation of cloud networks. As businesses continue to adopt cloud computing, it is essential for IT professionals and developers to understand the fundamental components and architectures that underpin cloud networking. By mastering these concepts, you will be better equipped to create efficient, scalable, and secure cloud-based solutions tailored to your organization's needs.

Cloud networks are comprised of various interconnected components that work together to enable communication and data exchange between resources in a cloud environment. These components are designed to provide flexibility, scalability, and security, which are vital for the successful deployment and management of applications and services in the cloud.

Throughout this section, we will explore these components and architectures in-depth, using real-world examples to illustrate their practical applications. For instance, you will learn how virtual networks and subnets can be utilized to isolate application tiers or to implement network access controls. Additionally, we will examine the role of gateways and routers in connecting different networks and facilitating data flow between them, using examples such as hybrid cloud implementations and multi-cloud strategies.

By the end of this section, you will have a comprehensive understanding of the various components that make up cloud networks and the architectural principles that guide their design. This knowledge will serve as a solid foundation for the rest of the module, enabling you to confidently design, implement, and manage cloud networks that meet your organization's requirements and objectives.

So, let's dive in and explore the world of cloud network components and architecture, and discover how they come together to create powerful, efficient, and secure cloud environments!

Virtual Networks

A virtual network, also known as a Virtual Private Cloud (VPC) in some cloud providers, is a software-defined network environment that operates within a cloud infrastructure. Virtual networks provide a secure and customizable layer of abstraction, allowing users to define and manage their network resources without the need to deal with the complexities of physical networking hardware.

Virtual networks are isolated from one another, which means that the resources within a virtual network can only communicate with other resources in the same network unless explicitly allowed through configured rules or gateways. This isolation enhances security and allows organizations to deploy applications and services in a controlled and protected environment.

Here are some key features of virtual networks:

- **Customizable IP Addressing:** Virtual networks offer the flexibility to define custom IP address ranges using CIDR notation. This allows organizations to design their network addressing schemes to suit their requirements, ensuring proper segmentation and efficient use of available IP address space.

- **Security Groups and Network ACLs:** Virtual networks support the use of security groups and network access control lists (ACLs) to manage and enforce access rules within the network. These features enable fine-grained control over network traffic, allowing organizations to restrict or allow communication between resources based on specific criteria, such as IP addresses, ports, or protocols.

- **Routing and Gateways:** Virtual networks provide built-in routing capabilities, allowing users to define custom routing rules and gateways for controlling traffic flow between different network segments or external networks. This enables scenarios such as connecting a virtual network to the public internet, an on-premises data center, or another virtual network within the same cloud environment.

- **Practical Example:** A software company wants to deploy a three-tier web application in a cloud environment. They create a virtual network with three subnets, one for each tier (web, application, and database). They define security groups and network ACLs to restrict traffic between the tiers, ensuring that only the necessary communication is allowed. Additionally, they configure an internet gateway to allow traffic from the public internet to reach the web tier, enabling users to access their web application.

In summary, virtual networks are a foundational component of cloud networking, providing a secure and customizable environment for deploying and managing cloud resources.

Understanding the concept of virtual networks will allow you effectively design and implement network architectures that meet your organization's requirements for security, scalability, and performance.

Subnets

Subnets are subdivisions of a virtual network that separate and organize resources into smaller, more manageable segments. Each subnet is assigned a unique IP address range within the larger address space of the virtual network. Subnets are essential for improving network performance, security, and organization by allowing administrators to define specific rules and policies for each segment.

Some key advantages of using subnets in a cloud environment include:

- **Network Segmentation:** Subnets allow for the logical separation of resources based on their function, application tier, or security requirements. This helps in managing access control and traffic flow more effectively.
- **Improved Performance:** By segmenting resources within a virtual network, administrators can reduce the amount of broadcast traffic and minimize network congestion, leading to better overall performance.
- **Enhanced Security:** Subnets can be used to isolate sensitive resources from other network segments, reducing the potential attack surface and limiting the blast radius in case of security incidents.
- **Scalability:** As the number of resources within a virtual network grows, subnets make it easier to manage and scale the network infrastructure by providing a clear organizational structure.
- **Practical Example:** A financial services company is deploying a multi-tier application in a cloud environment, consisting of a front-end web tier, a processing tier, and a database tier. The company decides to create three subnets within their virtual network, one for each tier. This allows them to apply specific security policies and access controls to each subnet, ensuring that the web tier can only communicate with the processing tier, and the processing tier can only communicate with the database tier. Additionally, they can isolate the database tier from public access, enhancing the overall security of their application.

To design and configure subnets, it is essential to understand the organization's requirements, such as the number of resources, their function, and the desired level of security. Administrators should also consider IP address allocation, ensuring that each subnet has a sufficient number of IP addresses for current and future needs.

In conclusion, subnets play a crucial role in cloud networking by dividing virtual networks into smaller, more manageable segments. By understanding how to design and configure subnets for different use cases, you can create efficient, secure, and organized network architectures that meet your organization's specific requirements.

Gateways and Routers

Gateways and routers play a pivotal role in cloud networking by connecting different networks and facilitating communication between them. They ensure that data is transmitted between networks in a secure and efficient manner, adhering to specific rules and policies defined by the administrators.

- **Gateways:** A gateway is a networking device that connects two or more networks with different address spaces and acts as an entry and exit point for network traffic. There are several types of gateways used in cloud networking:
    - **Internet Gateway:** An internet gateway allows resources within a virtual network to access the public internet and enables communication between the virtual network and external networks. It is crucial for providing public-facing services, such as web applications or APIs.
    - **VPN Gateway:** A VPN (Virtual Private Network) gateway establishes secure, encrypted connections between a virtual network and another network, such as an on-premises data center or another virtual network. VPN gateways are essential for creating hybrid cloud environments and ensuring secure communication between networks.
    - **NAT Gateway:** A Network Address Translation (NAT) gateway allows resources within a private subnet to access the internet without exposing their private IP addresses. NAT gateways help maintain the security

of internal resources while still allowing outbound internet access.

- **Routers:** A router is a networking device that forwards data packets between networks based on routing tables and routing protocols. In cloud networking, routers are responsible for directing traffic within a virtual network and between connected networks.

    - **Routing Tables:** Routing tables are used by routers to determine the best path for forwarding network traffic. They contain a set of rules that map destination IP address ranges to specific gateways or network interfaces. Administrators can configure custom routing tables to optimize data flow and enforce specific routing policies.

    - **Routing Protocols:** Routing protocols are used by routers to exchange routing information and maintain up-to-date routing tables. In cloud environments, dynamic routing protocols such as Border Gateway Protocol (BGP) are often used to ensure efficient and resilient routing between connected networks.

Practical Example: An e-commerce company has a hybrid cloud infrastructure, with their virtual network in the cloud connected to their on-premises data center through a VPN gateway. They also have an internet gateway to provide public access to their web application. To ensure efficient data flow between networks, they configure custom routing tables in their virtual network, directing traffic to the appropriate gateways based on the destination IP address.

In summary, gateways and routers are critical components of cloud networking, enabling secure and efficient communication between different networks. By understanding their roles and learning how to configure routing tables and gateways, you can create robust and scalable network architectures that effectively support your organization's cloud-based applications and services.

Load Balancers

Load balancers play a crucial role in cloud networking by distributing incoming network traffic across multiple servers or instances, optimizing resource utilization, and ensuring high availability and fault tolerance for cloud applications. By evenly distributing the load, they help prevent individual servers from becoming overwhelmed, leading to improved performance and reduced latency.

There are several load balancing techniques and cloud provider-specific implementations, including:

- **Algorithm-Based Load Balancing:** Load balancers use various algorithms to determine how to distribute incoming traffic among available servers. Some common algorithms include:
  - **Round Robin:** Traffic is distributed evenly across all servers in a sequential manner.
  - **Least Connections:** The server with the fewest active connections is selected to receive incoming traffic.
  - **IP Hash:** The client's IP address is used to determine which server will handle the request, ensuring that a specific client is always directed to the same server.
- **Layer 4 and Layer 7 Load Balancing:** Load balancers can operate at different layers of the OSI model, providing different levels of traffic management:
  - **Layer 4 (Transport Layer):** Load balancing is based on TCP or UDP connections, considering factors such as IP address and port number.

    - **Layer 7 (Application Layer):** Load balancing decisions are made based on the content of the request, such as HTTP headers or cookies, enabling more fine-grained traffic distribution.

- **Cloud Provider-Specific Implementations:** Major cloud providers offer their own managed load balancing services, which are designed to integrate seamlessly with their respective cloud environments:

    - **AWS Elastic Load Balancer (ELB):** Amazon Web Services (AWS) offers different types of load balancers, including Application Load Balancers (Layer 7), Network Load Balancers (Layer 4), and Classic Load Balancers (both Layer 4 and Layer 7).

    - **Google Cloud Load Balancer:** Google Cloud Platform (GCP) provides various load balancing options, such as HTTP(S) Load Balancing (Layer 7), TCP/SSL Proxy Load Balancing (Layer 4), and Internal Load Balancing (for traffic within the same VPC).

Practical Example: A media streaming company hosts their video content on a cluster of servers in the cloud. To ensure high availability and optimal performance, they use a cloud provider's managed load balancer to distribute incoming traffic across the cluster. By utilizing a Layer 7 load balancer, they can direct traffic based on the content type, sending video requests to specific servers optimized for video streaming.

In conclusion, load balancers are essential components of cloud networking, helping to distribute network traffic and ensure high availability and fault tolerance for cloud applications. By understanding different load balancing techniques and cloud provider-specific implementations, you can design and deploy effective load balancing strategies that enhance the performance, reliability, and scalability of your cloud-based applications and services.

Network Security

Network security is of paramount importance in cloud environments, as protecting your network from unauthorized access and potential threats is critical to maintaining the integrity and confidentiality of your data and applications. Various security measures can be employed to defend cloud networks, including network access control lists (ACLs), security groups, and firewall rules.

- **Network Access Control Lists (ACLs):** Network ACLs are stateless, layer 4 firewalls that control inbound and outbound traffic at the subnet level within a virtual network. They consist of a set of ordered rules that either allow or deny traffic based on criteria such as source or destination IP address, protocol, and port number. Network ACLs provide an additional layer of security by enforcing granular traffic filtering rules for each subnet.

- **Security Groups:** Security groups are stateful, layer 4 firewalls that control inbound and outbound traffic for individual resources, such as virtual machines or containers, within a virtual network. Security groups can be associated with multiple resources and are used to define rules that govern access to and from these resources. By using security groups, administrators can create fine-grained access policies that restrict communication between resources based on specific criteria, enhancing the overall security of the network.

- **Firewall Rules:** Firewall rules are an essential component of network security, enabling administrators to define specific rules that control the flow of traffic between resources, networks, and the internet. Firewall rules can be applied at various levels of the network stack, including layer 3 (IP) and layer 4 (transport), and can be used in conjunction with network ACLs and security groups to create a comprehensive security strategy.

Practical Example: A healthcare organization is deploying a patient management system in the cloud. To ensure the security of sensitive patient data, they implement multiple network security measures. They create network ACLs to enforce traffic filtering rules at the subnet level, restricting access to the database subnet to only the application subnet. Additionally, they configure security groups for the virtual machines running the application and database servers, allowing only the necessary communication between them. Finally, they implement firewall rules to restrict inbound traffic from the internet to only the web servers hosting the patient management system's user interface.

In conclusion, network security is a critical aspect of cloud networking that relies on a combination of ACLs, security groups, and firewall rules to protect resources from unauthorized access and potential threats. By understanding how these components work together, you can create a robust and secure network architecture that safeguards your cloud-based applications and data.

If you want to learn more about this point, I've published a book on Cybersecurity at Amazon (search: "Cybersecurity Certification Program for Game Changers").

DNS and Domain Management

The Domain Name System (DNS) is a vital component of the internet infrastructure and plays a crucial role in cloud networking. DNS translates human-readable domain names, such as www.example.com, into IP addresses that computers can understand, like 192.168.1.1. This translation process allows you to access websites and other online resources using easily recognizable names instead of having to remember complex IP addresses.

In cloud environments, you can leverage managed DNS services to set up and manage custom domain names and DNS records. Some popular cloud-based DNS services include Amazon Route 53 and Google Cloud DNS. These services provide a range of features that simplify domain management, such as domain registration, DNS record management, and traffic routing.

- **Domain Registration:** To use a custom domain name with your cloud resources, you need to register the domain through a domain registrar. Many cloud-based DNS services, including Amazon Route 53 and Google Cloud DNS, offer domain registration as part of their service offerings. By registering your domain with your chosen DNS service, you can manage all aspects of your domain and DNS configuration in one place.

- **DNS Record Management:** Once your domain is registered, you can create and manage various DNS records that define how traffic is routed to your cloud resources. Common DNS record types include:

  - **A Record:** Maps a domain name to an IPv4 address.

  - **AAAA Record:** Maps a domain name to an IPv6 address.

  - **CNAME Record:** Creates an alias for another domain name, allowing you to direct traffic from one domain to another.

  - **MX Record:** Specifies the mail server responsible for handling email for your domain.

- **Traffic Routing:** Cloud-based DNS services also offer advanced traffic routing features that allow you to optimize the flow of traffic to your resources.

  For example, you can implement latency-based routing to direct users to the closest available resource, or use weighted routing to distribute traffic across multiple resources based on specific weights.

Practical Example: Suppose you are deploying a web application in the cloud and want to use a custom domain name for your application. First, you would register your desired domain name using a DNS service like Amazon Route 53 or Google Cloud DNS. Next, you would create an A record to map your domain name to the IP address of your cloud-based web server. If you have additional resources, such as an email server or an API, you can create MX or CNAME records to route traffic accordingly.

In summary, understanding DNS and domain management is essential for creating user-friendly and easily accessible cloud-based applications and services. By learning how to set up and manage custom domain names and DNS records using cloud-based DNS services, you can ensure that your online resources are easily discoverable and efficiently routed to provide the best possible user experience.

Content Delivery Networks (CDNs)

Content Delivery Networks (CDNs) are essential components of modern internet infrastructure that play a crucial role in optimizing the delivery of content to end-users worldwide. By caching and distributing content across multiple edge locations, CDNs can significantly reduce latency and improve the overall user experience for your cloud-based applications and services.

- **Caching Content:** CDNs cache static content, such as images, stylesheets, and JavaScript files, at strategically placed edge locations, which are geographically closer to end-users. When a user requests content from your application, the CDN checks if the content is already cached at an edge location. If the content is available, the CDN delivers it directly to the user, significantly reducing the time it takes for the content to travel from your origin server to the user's device.

- **Distributing Content:** CDNs automatically distribute your content to multiple edge locations, ensuring that it is available to users regardless of their geographic location. By storing your content closer to your users, CDNs can reduce the impact of network congestion, improve content delivery speeds, and enhance the overall performance of your application.
- **Reducing Latency:** Latency is the time it takes for data to travel between the server and the client. High latency can negatively affect the user experience, particularly for applications that rely on real-time data or streaming media. CDNs help reduce latency by delivering content from edge locations that are geographically closer to end-users, minimizing the distance data must travel.

Practical Example: Imagine you have a cloud-based e-commerce website that serves customers worldwide. To ensure a fast and responsive user experience, you decide to leverage a CDN to optimize content delivery. You configure the CDN to cache your website's static assets, such as product images and stylesheets, at multiple edge locations around the globe. When a user visits your website, the CDN delivers the cached assets from the edge location closest to the user, significantly reducing latency and improving the overall user experience.

In conclusion, Content Delivery Networks are a powerful tool for optimizing the delivery of content to users across the globe. By understanding how CDNs cache and distribute content across multiple edge locations, you can harness their potential to reduce latency and enhance the user experience for your cloud-based applications and services.

Hybrid and Multi-Cloud Networking

As you advance in your cloud networking journey, you may encounter scenarios where you need to connect your on-premises data center to a cloud environment or integrate services from multiple cloud providers. These advanced networking concepts, known as hybrid networking and multi-cloud networking, present unique challenges and benefits.

- **Hybrid Networking:** Hybrid networking refers to the connection of on-premises data centers to cloud environments, enabling you to extend your existing infrastructure and take advantage of the scalability and flexibility offered by the cloud. This approach allows you to maintain sensitive data on-premises while utilizing cloud resources for processing and analytics. Key tools and techniques for implementing hybrid networking include:
    - **Virtual Private Networks (VPNs):** Establish secure, encrypted connections between your on-premises data center and cloud environment.
    - **Direct Connect or ExpressRoute:** Create dedicated, high-speed connections between your on-premises data center and cloud provider, bypassing the public internet for increased security and performance.
- **Multi-Cloud Networking:** Multi-cloud networking involves integrating services from multiple cloud providers, allowing you to leverage the unique offerings and capabilities of each provider to create a more resilient and diverse infrastructure. This approach can help you avoid vendor lock-in, optimize costs, and take advantage of best-of-breed services. To implement multi-cloud networking, you need to:
    - Understand each cloud provider's networking constructs and services, as well as how they can be interconnected.
    - Leverage tools such as Terraform or Ansible to automate the provisioning and management of resources across multiple cloud providers.

Practical Example: Imagine you are responsible for the IT infrastructure of a large organization that operates both an on-premises data center and cloud-based services. To enhance your infrastructure's resilience and flexibility, you decide to adopt a hybrid and multi-cloud networking approach.

To connect your on-premises data center to your cloud environment, you establish a VPN connection, creating a secure and encrypted communication channel. Additionally, you use Direct Connect to create a high-speed, dedicated connection between your data center and cloud provider for increased performance.

For multi-cloud networking, you choose to deploy your application across two cloud providers, leveraging each provider's unique strengths. You use Terraform to automate the provisioning of resources and manage your multi-cloud infrastructure consistently.

Understanding the challenges and benefits of hybrid and multi-cloud networking, as well as the tools and techniques available for implementing these approaches, will allow you create a more robust, scalable, and resilient IT infrastructure that meets your organization's unique needs.

Implementing a Cloud-Based Virtual Private Network (VPN)

In section you will delve into the world of VPNs and their essential role in cloud networking. VPNs are a key security measure for protecting data and ensuring secure communication between on-premises infrastructure and cloud environments, as well as between different cloud environments. By creating an encrypted and private tunnel over the public internet, VPNs help safeguard your data from eavesdropping, tampering, and unauthorized access.

We will explore the various types of VPNs available, their use cases, and the process of setting up and managing a VPN in a cloud environment. For instance, you may want to establish a VPN connection between your on-premises data center and a cloud provider to extend your existing infrastructure securely. Alternatively, you might need to connect two cloud environments from different providers to enable seamless communication and data transfer between them.

To illustrate the practical implementation of a cloud-based VPN, consider a scenario where your organization has an on-premises data center and wants to utilize cloud-based resources for additional processing power and storage. By setting up a VPN connection between your data center and the cloud environment, you can maintain a secure link for data transfer and communication while reaping the benefits of cloud computing.

Throughout this section, you will gain valuable insights into implementing a cloud-based VPN solution and learn how to configure, monitor, and maintain VPN connections to ensure the security and integrity of your data and communications in a cloud environment.

Types of Cloud-Based VPNs

In the realm of cloud-based VPNs, there are two primary types that you need to be familiar with: site-to-site VPNs and remote access VPNs. Each type serves a specific purpose and caters to different use cases, so understanding their distinct features and functions will help you select the most suitable VPN solution for your needs.

- **Site-to-Site VPNs:** Site-to-site VPNs establish secure connections between two networks, typically connecting an on-premises network to a cloud environment, or interconnecting networks across multiple cloud providers. These VPNs are ideal for situations where you need to securely transmit data between different locations or extend your existing infrastructure to the cloud. Site-to-site VPNs utilize VPN gateways to encrypt and encapsulate data, creating a secure tunnel for communication between networks.

  Practical Example: Suppose you have an on-premises data center and decide to utilize a cloud provider for additional storage and processing capabilities. By setting up a site-to-site VPN between your data center and the cloud environment, you can securely transfer data and maintain seamless communication between the two networks.

- **Remote Access VPNs:** Remote access VPNs enable individual users to securely access a private network, such as your organization's internal network, from remote locations over the internet. This type of VPN is useful when you have employees who work remotely or need to access sensitive data and resources from outside the office. Remote access VPNs rely on a VPN client installed on the user's device, which establishes an encrypted connection to a VPN server within the private network.

  Practical Example: Consider a scenario where you are working from home and need to access your organization's internal applications and resources. By using a remote access VPN, you can securely connect to your organization's network and access the required resources as if you were physically present in the office.

In conclusion, understanding the differences between site-to-site and remote access VPNs is crucial for choosing the right VPN solution for your cloud environment. By considering your organization's specific needs and use cases, you can select the most appropriate VPN type to ensure secure communication and data transmission between networks or remote users.

Establishing VPN Connections

Setting up a VPN connection is a crucial step in securing communication between your on-premises infrastructure and a cloud environment or between multiple cloud environments. This process typically involves configuring VPN gateways and customer gateways to create a secure, encrypted tunnel for data transmission.

- **VPN Gateway:** A VPN gateway is a network device that serves as an endpoint for VPN connections. In a cloud environment, the VPN gateway is managed by the cloud provider and is responsible for encrypting and encapsulating data before it's transmitted over the public internet. You'll need to configure the VPN gateway in your cloud provider's management console, specifying the required settings such as the IP address of the customer gateway and the encryption protocols to be used.

- **Customer Gateway:** The customer gateway represents the endpoint of the VPN connection in your on-premises network or another cloud environment. This can be a physical or virtual device, such as a router or firewall, that supports VPN functionality. You'll need to configure the customer gateway with the appropriate settings, including the IP address of the VPN gateway and the encryption protocols to be used.

Practical Example: Let's say you want to establish a VPN connection between your on-premises data center and a cloud environment hosted on AWS. Here's an overview of the steps you would take:

- In the AWS Management Console, create a Virtual Private Gateway and attach it to your Virtual Private Cloud (VPC).
- Configure the Customer Gateway in the AWS Management Console, providing the public IP address of your on-premises VPN device.
- Create a Site-to-Site VPN Connection, specifying the Virtual Private Gateway and Customer Gateway created in the previous steps.
- Download the VPN configuration file from AWS and apply the settings to your on-premises VPN device, ensuring that the encryption protocols and other settings match those configured in the AWS Management Console.

Following these steps and configuring the VPN gateways and customer gateways correctly will allow you establish a secure VPN connection between your on-premises infrastructure and a cloud environment or between multiple cloud environments. This will enable you to securely transmit data and maintain seamless communication between networks while leveraging the benefits of cloud computing.

VPN Encryption and Security

One of the critical aspects of VPN connections is the encryption and security measures used to protect your data during transmission. These measures ensure the confidentiality, integrity, and authentication of your data, making VPNs a reliable solution for securing communication between networks. Two primary encryption protocols are widely used in VPN connections: IPsec and SSL/TLS.

- **IPsec (Internet Protocol Security):** IPsec is a popular encryption protocol used in site-to-site VPNs and some remote access VPNs. It operates at the network layer, providing end-to-end security for data packets transmitted over the internet. IPsec employs various cryptographic algorithms and protocols, such as Encapsulating Security Payload (ESP) for data confidentiality and integrity, and Internet Key Exchange (IKE) for secure key exchange and authentication.

  Practical Example: If you set up a site-to-site VPN connection between your on-premises data center and a cloud environment, you might choose to use IPsec as the encryption protocol. This will ensure that your data is securely transmitted over the internet, safeguarding it from eavesdropping and tampering.

- **SSL/TLS (Secure Sockets Layer/Transport Layer Security):** SSL/TLS is a widely-used encryption protocol commonly employed in remote access VPNs and web-based applications. It operates at the transport layer, creating a secure channel for data transmission between a client and a server. SSL/TLS provides data confidentiality, integrity, and authentication through the use of digital certificates and asymmetric cryptography.

  Practical Example: If you need to remotely access your organization's internal network from a public network, you might use a remote access VPN with SSL/TLS encryption. By establishing a secure connection between your device and the VPN server, you can access sensitive data and resources without exposing them to potential threats on the public internet.

In summary, understanding the differences between IPsec and SSL/TLS encryption protocols is essential for ensuring the security of your VPN connections. By selecting the appropriate encryption protocol based on your use case, you can effectively protect the confidentiality, integrity, and authentication of your data while transmitting it over VPN connections.

Monitoring and Troubleshooting VPNs

Proactively monitoring and troubleshooting your VPN connections are essential aspects of maintaining secure and reliable communication between networks. By keeping an eye on the performance and status of your cloud-based VPNs, you can identify potential issues and resolve them before they escalate into more significant problems.

- **Monitoring VPNs:** Monitoring the performance and status of your VPN connections involves tracking various metrics and indicators, such as the number of active connections, latency, and data throughput. Most cloud providers offer built-in monitoring tools that can help you keep track of these metrics. For example, AWS provides CloudWatch, while Azure offers Azure Monitor, and Google Cloud Platform has Stackdriver Monitoring.

  Practical Example: You can set up monitoring and alerts in CloudWatch to notify you when specific thresholds are exceeded, such as unusually high latency or low data throughput. By regularly checking these metrics and responding to alerts, you can ensure the optimal performance of your VPN connections and address any potential issues before they impact your operations.

- **Troubleshooting VPNs:** When issues arise during the setup or maintenance of your VPN connections, it's crucial to identify the root cause and implement the necessary fixes. Common VPN issues include connection failures, incorrect configurations, and performance degradation. To troubleshoot these problems, you can consult logs, use diagnostic tools provided by the cloud provider, and perform tests to isolate the issue.

Practical Example: Suppose you encounter a connection failure between your on-premises network and a cloud environment. To troubleshoot the issue, you might start by checking the logs on both the customer gateway and the VPN gateway for any errors or inconsistencies. Next, verify the configurations on both sides to ensure they match, and check the connectivity between the gateways. By following a systematic approach to troubleshooting, you can pinpoint the root cause and resolve the issue, ensuring the seamless operation of your VPN connections.

In conclusion, monitoring and troubleshooting are essential aspects of managing cloud-based VPNs. By proactively tracking the performance and status of your VPN connections and addressing any issues that arise, you can maintain secure and reliable communication between your on-premises infrastructure and cloud environments or between multiple cloud environments.

Cloud Provider VPN Services

To implement a cloud-based VPN, it's essential to understand the VPN services offered by major cloud providers. Each provider has its unique features and capabilities, which may affect your decision when selecting the best solution for your organization. Below, you'll find an overview of the VPN offerings from AWS, Azure, and Google Cloud.

- **AWS Virtual Private Gateway:** AWS offers a Virtual Private Gateway (VGW) as a solution for creating site-to-site VPN connections between your on-premises network and your Amazon VPCs. AWS VGW supports both IPsec and SSL/TLS VPNs and provides features such as high availability through multiple VPN tunnels, dynamic routing, and integration with AWS Direct Connect for improved performance.

  Practical Example: If your organization uses AWS as your cloud provider, you can set up a site-to-site VPN connection between your on-premises network and an

Amazon VPC using the AWS VGW. This connection will enable secure communication between your on-premises resources and your cloud-based infrastructure.

- **Azure VPN Gateway:** Azure VPN Gateway is a managed VPN service that enables you to create site-to-site and remote access VPN connections between your on-premises network and Azure Virtual Networks (VNets). Azure VPN Gateway supports IPsec and IKEv2 VPN protocols, providing features such as point-to-site VPN connections, forced tunneling, and BGP routing.

  Practical Example: If you're using Azure as your cloud provider, you can establish a site-to-site VPN connection between your on-premises network and an Azure VNet using the Azure VPN Gateway. This secure connection allows you to access your Azure resources while maintaining a high level of security and control.

- **Google Cloud VPN:** Google Cloud VPN is a fully managed VPN service that enables you to create site-to-site VPN connections between your on-premises network and Google Cloud VPCs. Google Cloud VPN supports IPsec VPNs and offers features such as high availability, dynamic routing, and integration with Google Cloud Interconnect for improved performance.

  Practical Example: If your organization utilizes Google Cloud Platform, you can establish a site-to-site VPN connection between your on-premises network and a Google Cloud VPC using Google Cloud VPN. This connection ensures secure data transmission between your on-premises infrastructure and your cloud-based resources.

In summary, understanding the differences and unique features of each cloud provider's VPN services is crucial when choosing the best solution for your organization's needs. By familiarizing yourself with the offerings from AWS, Azure, and Google Cloud, you can make an informed decision and implement a secure, reliable VPN connection that meets your specific requirements.

Practical: Configuring Cloud Networking Services

In this section, "3.3 Practical: Configuring Cloud Networking Services," you will gain hands-on experience with the essential aspects of cloud networking, focusing on the configuration and management of various networking components within major cloud providers. Through a series of practical examples and guided exercises, you will learn how to set up and fine-tune different network components to optimize performance, security, and reliability in your cloud infrastructure.

You will have the opportunity to work with virtual networks, subnets, gateways, and routers, as well as configuring load balancers and implementing network security measures. By working with real-world scenarios, you will gain a deeper understanding of how these components interact and how to configure them to meet the specific requirements of your organization.

Whether you are using AWS, Azure, Google Cloud, or another cloud provider, the practical exercises in this section will provide you with valuable skills and insights to help you effectively configure and manage cloud networking services for your projects. Armed with this knowledge, you will be better equipped to make informed decisions and ensure the success of your cloud deployments.

Creating Virtual Networks

In this exercise, you'll learn how to create and configure virtual networks using cloud provider management consoles or command-line interfaces. You'll also understand how to set up network configurations, such as IP address ranges and DNS settings.

A virtual network is an isolated, software-defined network environment within a cloud infrastructure that allows you to create custom network configurations, secure communication between resources, and manage network traffic. As you create a virtual network, it's essential to properly configure its settings, such as IP address ranges and DNS settings, to ensure the smooth operation of your cloud resources.

Practical Example: Suppose you are using Azure as your cloud provider. To create a virtual network, you can follow these steps:

1. Sign in to the Azure portal and navigate to the "Virtual Networks" section.
2. Click on "Add" to create a new virtual network.
3. Choose a subscription and a resource group for your virtual network.
4. Enter a name for the virtual network and specify the region where it will be deployed.
5. Configure the IP address range by providing an address space in CIDR notation (e.g., 10.0.0.0/16).
6. Set up a default subnet by entering a name and subnet address range (e.g., 10.0.1.0/24).
7. Configure DNS settings by selecting the desired DNS servers (either custom or default Azure-provided DNS).
8. Review your settings and click "Create" to deploy the virtual network.

With your virtual network now created and configured, you can connect and manage your cloud resources more effectively. By understanding the process of creating and configuring virtual networks, you'll be better equipped to manage your cloud infrastructure and optimize network performance, security, and organization.

Configuring Subnets

In this exercise, you'll learn how to create and manage subnets within virtual networks, assign IP address ranges, and implement network access control to improve network performance, security, and organization.

Subnets are smaller, manageable segments within a virtual network that help you to logically separate resources and control network traffic. Creating subnets allows you to better organize your cloud resources, implement network access control, and enhance security.

Practical Example: Suppose you are using AWS as your cloud provider and have already created a virtual private cloud (VPC). To create a subnet within your VPC, you can follow these steps:

1. Sign in to the AWS Management Console and navigate to the "VPC" section.
2. Click on "Subnets" in the left-hand menu, then click "Create subnet."
3. Select the VPC in which you want to create the subnet.
4. Enter a name and description for the subnet.
5. Choose an availability zone for your subnet or let AWS select one for you.
6. Specify the IPv4 CIDR block for the subnet (e.g., 10.0.1.0/24). Make sure it falls within the address range of your VPC.
7. Click "Create subnet" to deploy the new subnet within your VPC.

After creating the subnet, you can configure network access control by creating and associating security groups and network access control lists (ACLs) with your subnet. This allows you to define inbound and outbound traffic rules, effectively controlling which resources can communicate with one another.

By understanding how to create and manage subnets within virtual networks, you will be able to optimize your cloud infrastructure's performance, security, and organization, ensuring a more efficient and reliable environment for your cloud resources.

Setting Up Gateways and Routing

In this exercise, you'll gain practical experience in configuring VPN gateways, Internet gateways, and routing tables for efficient data flow between networks, using cloud provider-specific tools and services.

Gateways and routing tables are essential components of cloud networking, as they facilitate communication between different networks and ensure data flow between resources is efficient and secure.

Practical Example: Suppose you are using Google Cloud as your cloud provider and want to set up an Internet gateway and routing for your virtual private cloud (VPC) network. To do this, follow these steps:

1. Sign in to the Google Cloud Console and navigate to the "VPC networks" section.
2. Click on the name of the VPC network you want to configure.
3. To create an Internet gateway, go to the "Firewall" section in the left-hand menu and click "Create firewall rule."
4. Enter a name and description for the rule, and set the priority (lower values have higher priority).
5. Define the source IP ranges, such as 0.0.0.0/0 for all IP addresses.

6. Choose the allowed protocols and ports for inbound traffic, for example, TCP:80,443 for HTTP and HTTPS.

7. Apply the rule to the instances or subnets in your VPC network using target tags or service accounts.

8. Click "Create" to deploy the firewall rule and allow incoming traffic from the Internet.

Next, configure the routing table:

9. In the "VPC networks" section, click on "Routes" in the left-hand menu, then click "Create route."

10. Select the VPC network for which you want to create the route.

11. Enter a name and description for the route.

12. Specify the destination IP range (e.g., 0.0.0.0/0 for all IP addresses).

13. Choose the next hop, which is where the traffic will be sent. For an Internet gateway, select "Default Internet gateway."

14. Click "Create" to add the route to your VPC network's routing table.

By setting up gateways and routing, you will ensure efficient data flow between your cloud resources and the Internet or other networks. With this practical experience, you'll be better prepared to manage your cloud infrastructure and optimize network performance.

Implementing Load Balancing

In this exercise, you'll learn how to set up and configure cloud-based load balancers to distribute network traffic across multiple servers or instances, ensuring high availability and fault tolerance for cloud applications.

Load balancers play a crucial role in managing and distributing network traffic to maintain optimal performance and prevent any single server or instance from being overwhelmed.

Practical Example: Suppose you are using Azure as your cloud provider and want to create a load balancer to distribute traffic among virtual machines (VMs) in your resource group. To set up an Azure Load Balancer, follow these steps:

1. Sign in to the Azure portal and navigate to the "Load Balancers" section.
2. Click "Add" to create a new load balancer.
3. Choose a subscription and resource group for your load balancer.
4. Enter a name for the load balancer and select the region.
5. Choose the type of load balancer: public (accessible from the Internet) or internal (accessible only within your virtual network).
6. Select the SKU: Basic or Standard. Standard load balancers offer more features and better performance.
7. Click "Review + create" and then "Create" to deploy the load balancer.

Next, configure the backend pool:

8. In the "Load Balancers" section, click on the name of your newly created load balancer.
9. In the left-hand menu, click "Backend pools" and then "Add."
10. Enter a name for the backend pool and choose the virtual network containing your VMs.
11. Add VMs or instances to the backend pool by selecting them from the list.

12. Click "Add" to save the backend pool configuration.

Finally, configure the load balancing rules:

13. In the left-hand menu, click "Load balancing rules" and then "Add."
14. Enter a name for the rule and select the frontend IP configuration, backend pool, and health probe.
15. Choose the protocol (TCP or UDP), and set the frontend and backend port numbers.
16. Configure session persistence and idle timeout settings as needed.
17. Click "Add" to create the load balancing rule.

Once you've completed these steps, your Azure Load Balancer will distribute incoming traffic across the VMs in your backend pool, ensuring high availability and fault tolerance for your cloud applications. By learning how to implement load balancing in cloud environments, you'll be better equipped to manage and optimize your cloud infrastructure.

Securing Cloud Networks

In this exercise, you'll gain hands-on experience in applying network security measures, such as network access control lists (ACLs), security groups, and firewall rules, to protect your cloud resources from unauthorized access and potential threats.

Securing your cloud network is crucial to safeguard sensitive data and maintain the integrity of your infrastructure. By implementing robust security measures, you can prevent unauthorized access and mitigate potential cyber threats.

Practical Example: Suppose you are using AWS as your cloud provider and want to create and configure a security group to protect your EC2 instances. To set up an AWS security group, follow these steps:

1. Sign in to the AWS Management Console and navigate to the "EC2" section.
2. In the left-hand menu, click "Security Groups" under the "Network & Security" category.
3. Click "Create security group" to start the process.
4. Enter a name and description for your security group, and choose the Virtual Private Cloud (VPC) in which your instances reside.
5. Click "Create security group."
6. Next, configure inbound and outbound rules for your security group:
7. In the "Security Groups" section, select the newly created security group and click on the "Inbound rules" tab.
    1. Click "Edit inbound rules" and then "Add rule."
8. Choose the type of traffic you want to allow (e.g., HTTP, HTTPS, SSH), specify the source (e.g., custom IP range, another security group), and add a description for the rule.
9. Click "Save rules" to apply your inbound rules.
10. Switch to the "Outbound rules" tab and click "Edit outbound rules."
11. Add rules for the types of outbound traffic you want to allow, specifying the destination and adding a description for each rule.
12. Click "Save rules" to apply your outbound rules.

After completing these steps, your AWS security group will be configured to allow only the specified traffic to and from your EC2 instances, effectively protecting your cloud resources from unauthorized access and potential threats.

Learning how to secure cloud networks with ACLs, security groups, and firewall rules, will allow you to be better prepared to manage and maintain a secure cloud infrastructure.

Module 3 Exam Questions

1. What is the primary purpose of a virtual network in a cloud infrastructure?

a) Manage security groups

b) Provide a software-defined, isolated network environment

c) Implement content delivery networks

d) Deploy serverless applications

2. Which of the following is a benefit of dividing a virtual network into subnets?

a) Increased latency

b) Improved network performance

c) Reduced security

d) Simplified routing tables

3. What is the primary role of a gateway in cloud networking?

a) Managing subnets

b) Configuring DNS records

c) Connecting different networks

d) Monitoring network performance

4. Which cloud service is used to distribute network traffic across multiple servers or instances?

a) Content delivery network

b) Load balancer

c) Virtual private network

d) Domain name system

5. What does a network access control list (ACL) help to achieve?

a) Managing domain names

b) Reducing latency

c) Protecting cloud resources

d) Configuring routing tables

6. Which cloud service translates human-readable domain names into IP addresses?

a) Load balancer

b) Content delivery network

c) Virtual private network

d) Domain name system

7. What is the primary purpose of a content delivery network (CDN)?

a) Optimizing content delivery to end-users

b) Managing domain names

c) Connecting different networks

d) Implementing network access control

8. In a hybrid cloud networking setup, what does the infrastructure typically consist of?

a) On-premises data center and cloud environment

b) Multiple cloud providers

c) Only cloud environments

d) None of the above

9. Which type of VPN is suitable for connecting an on-premises infrastructure to a cloud environment?

a) Remote access VPN

b) Site-to-site VPN

c) Both a and b

d) None of the above

10. What is a key security measure used in VPNs to ensure the confidentiality of data transmitted over the connection?

a) Load balancing

b) Routing

c) Encryption

d) Subnetting

11. Which cloud provider offers the Elastic Load Balancer service?

a) AWS

b) Azure

c) Google Cloud

d) IBM Cloud

12. In which section of the AWS Management Console can you find Security Groups?

a) EC2

b) S3

c) Lambda

d) RDS

13. What is the main purpose of a security group in a cloud environment?

a) Managing domain names

b) Routing traffic

c) Protecting cloud resources

d) Reducing latency

14. Which protocol is commonly used for securing VPN connections?

a) HTTP

b) IPsec

c) FTP

d) SMTP

15. Which cloud provider offers the Azure VPN Gateway service?

a) AWS

b) Azure

c) Google Cloud

d) IBM Cloud

16. What is the primary benefit of a multi-cloud networking strategy?

a) Simplified network management

b) Reduced costs

c) Integration of multiple cloud providers' services

d) Increased latency

17. Which of the following is a cloud-based DNS service?

a) Amazon Route 53

b) Azure Functions

c) Google Cloud Load Balancer

d) AWS Elastic Beanstalk

18. Which of the following is a benefit of implementing a load balancer?

a) Ensuring high availability

b) Increasing network latency

c) Reducing network security

d) Simplifying network management

19. What type of VPN is suitable for remote users to securely access corporate resources?

a) Remote access VPN

b) Site-to-site VPN

c) Both a and b

d) None of the above

20. In a cloud-based virtual private network (VPN), what is the purpose of a customer gateway?

a) It manages domain names

b) It represents the on-premises side of a VPN connection

c) It optimizes content delivery

d) It distributes network traffic across multiple instances

Answers:

1. b
2. c
3. a
4. d
5. a
6. d
7. b
8. c
9. d
10. b
11. c
12. a
13. a
14. d
15. b
16. c
17. b
18. a
19. d
20. b

Module 4. Cloud Security and Compliance

Introduction to Cloud Security and Compliance

In today's digital landscape, ensuring the security and compliance of cloud-based systems is of paramount importance. As organizations continue to migrate their infrastructure and applications to the cloud, they face numerous challenges, including safeguarding sensitive data, managing user access, and adhering to regulatory requirements. This module aims to equip you with the knowledge and skills necessary to implement robust security measures and maintain compliance in your cloud environment.

Throughout this module, we will explore the various aspects of cloud security and compliance, drawing on real-world examples and case studies to illustrate key concepts and best practices. We will delve into topics such as data protection, secure network design, and risk management. By understanding these principles, you will be better prepared to create and maintain a secure cloud ecosystem.

To help you grasp the practical aspects of cloud security, we will also provide hands-on exercises and demonstrations that showcase how to implement security measures using popular cloud service providers like Amazon Web Services (AWS), Microsoft Azure, and Google Cloud Platform (GCP). These practical examples will give you the opportunity to apply what you've learned and build confidence in your ability to secure your own cloud infrastructure.

By the end of this module, you will have a comprehensive understanding of cloud security and compliance and the practical know-how to implement effective security measures. Armed with this knowledge, you will be well-equipped to protect your organization's valuable assets and maintain a secure, compliant cloud environment.

Let's dive into the exciting world of cloud security and compliance and start building the foundation for a secure cloud infrastructure.

Note: If you want to learn even more about this topic, I've published a book on Cybersecurity at Amazon (search: "Cybersecurity Certification Program for Game Changers").

Understanding Cloud Security Best Practices

As organizations increasingly rely on cloud-based solutions, it is essential to ensure the security of their cloud infrastructure and applications. This section aims to provide you with a solid foundation in cloud security best practices, enabling you to identify potential vulnerabilities and implement robust security measures to protect your organization's valuable assets.

Throughout this section, we will delve into the core principles of cloud security, discussing various concepts and techniques that contribute to a secure and resilient cloud environment. We will examine the shared responsibility model, which highlights the division of security responsibilities between cloud service providers and their customers. By understanding this model, you will be better prepared to fulfill your organization's security obligations in the cloud.

To help you grasp the practical aspects of cloud security, we will also draw on real-world examples and case studies, illustrating how organizations have successfully implemented security best practices to safeguard their cloud infrastructure. These examples will provide valuable insights into the challenges faced by organizations of different sizes and industries, allowing you to learn from their experiences and apply the lessons learned to your own cloud security strategy.

By the end of this section, you will have a comprehensive understanding of cloud security best practices and the knowledge required to assess your organization's cloud security posture. Armed with this information, you can take the necessary steps to protect your cloud environment and maintain a secure, compliant infrastructure.

Let's begin our journey into the world of cloud security best practices, laying the groundwork for a secure and resilient cloud ecosystem.

Data Encryption

In the world of cloud computing, data encryption is a crucial component of your security strategy. It ensures that your sensitive information remains secure and inaccessible to unauthorized users, both when stored (at rest) and when transmitted between systems (in transit). Let's explore the importance of data encryption, various encryption algorithms, key management strategies, and best practices for securing data in the cloud.

Encryption algorithms are mathematical functions that convert plaintext data into unreadable ciphertext, using an encryption key. Some widely-used encryption algorithms include Advanced Encryption Standard (AES), RSA, and Triple Data Encryption Standard (3DES). Selecting the right encryption algorithm for your use case is essential for providing a strong security foundation.

Key management refers to the process of creating, distributing, storing, and retiring encryption keys. Effective key management ensures that only authorized users can access the encrypted data. Cloud service providers often offer key management services, such as AWS Key Management Service (KMS), Azure Key Vault, or Google Cloud KMS, which allow you to manage encryption keys in a secure and compliant manner.

Here are some best practices for securing data in the cloud using encryption:

- **Encrypt data at rest:** Ensure that all sensitive data stored in the cloud is encrypted. This includes databases, file storage, and backups. Most cloud providers offer built-in encryption options for their storage services, making it easy to enable encryption with minimal configuration.

- **Encrypt data in transit:** Protect data transmitted over networks by using encryption protocols like Transport Layer Security (TLS) or Secure Shell (SSH). These protocols ensure that data is encrypted before it leaves your system and remains encrypted until it reaches its destination.

- **Use strong encryption algorithms:** Choose encryption algorithms that are widely-accepted and have been thoroughly tested by the cryptographic community, such as AES-256 or RSA-2048.

- **Implement proper key management:** Keep encryption keys secure by storing them in dedicated key management services or hardware security modules (HSMs). Regularly rotate encryption keys and revoke access to keys when they are no longer needed.

Following these best practices will allow you significantly enhance the security of your data in the cloud, minimizing the risk of unauthorized access and ensuring compliance with industry regulations. Implementing data encryption is an essential step towards building a robust and secure cloud environment.

Secure Network Design

Designing a secure cloud network is essential for protecting your sensitive data and resources from potential threats. By implementing appropriate security measures, such as security groups, access control lists (ACLs), virtual private clouds (VPCs), and others, you can create a robust and resilient network environment. In this section, we'll explore various network security measures and learn how to isolate resources, segment networks, and use VPNs or Direct Connect for secure communication.

- **Security Groups:** Security groups are virtual firewalls that control inbound and outbound traffic to your cloud resources. They allow you to define rules specifying which traffic is allowed or denied based on IP addresses, protocols, and ports. When configuring security groups, ensure that you follow the principle of least privilege, allowing only the necessary traffic and blocking everything else by default.

- **Access Control Lists (ACLs):** ACLs provide an additional layer of security by controlling traffic at the subnet level within your VPC. They allow you to create granular rules to permit or deny traffic based on source and destination IP addresses and ports. Combining ACLs with security groups helps you create a more secure network environment.

- **Virtual Private Clouds (VPCs):** VPCs enable you to create isolated network environments within your cloud infrastructure. They allow you to define your own IP address range, subnets, and routing tables, ensuring that your resources are logically separated from other users within the same cloud provider. By using VPCs, you can effectively segment your network and control access to your resources.

- **Network Segmentation:** Network segmentation involves dividing your network into smaller, isolated segments to limit the potential attack surface. You can achieve this by creating separate VPCs or subnets for different applications, environments, or departments. This helps you restrict access to sensitive resources and contain potential security incidents within a specific network segment.

- **Secure Communication:** When connecting your on-premises data center to your cloud environment or when transmitting data between different cloud regions, it's essential to protect the communication channels. You can use Virtual Private Networks (VPNs) or Direct Connect to establish secure, encrypted connections between your networks. VPNs create a secure tunnel over the public internet, while Direct Connect provides a dedicated, private connection between your on-premises network and your cloud infrastructure.

Implementing these network security measures willl allow you design a secure cloud network that effectively protects your resources and data from unauthorized access and potential threats. A well-designed network architecture is the cornerstone of a secure and resilient cloud environment.

Identity and Access Management (IAM)

Managing user access to your cloud resources is a critical aspect of ensuring your cloud environment's security. Identity and Access Management (IAM) is the process of defining, enforcing, and auditing who has access to which resources and services within your cloud infrastructure. In this section, we'll discuss IAM best practices, such as least privilege, role-based access control (RBAC), and multi-factor authentication (MFA), and explore how to create and manage user accounts, groups, and roles effectively.

- **Least Privilege:** The principle of least privilege states that users should only have the minimum access necessary to perform their job functions. This approach helps minimize the risk of unauthorized access and potential damage caused by compromised accounts. When configuring access permissions, ensure that you grant only the required privileges and nothing more.

- **Role-Based Access Control (RBAC):** RBAC is a method of managing user access based on predefined roles that are assigned to users based on their job responsibilities. Each role has a specific set of permissions that determine what actions users can perform within the cloud environment. By implementing RBAC, you can simplify the process of managing access permissions and ensure that users only have access to the resources they need.

- **Multi-Factor Authentication (MFA):** MFA adds an additional layer of security to the authentication process by requiring users to provide two or more forms of identification before accessing the cloud environment. This can include something the user knows (e.g., a password), something the user has (e.g., a hardware token), or something the user is (e.g., a fingerprint). Enabling MFA can significantly reduce the risk of unauthorized access due to compromised credentials.

- **User Accounts and Groups:** Creating and managing user accounts is the first step in implementing IAM within your cloud environment. Assign users to groups based on their job functions or department, and grant access permissions at the group level. This approach simplifies the process of managing access and ensures that users have the appropriate permissions based on their roles within the organization.

- **Roles and Policies:** Create roles with specific permissions and assign them to users or groups as needed. Use policies to define the permissions associated with each role, specifying which actions can be performed on which resources. This allows you to granularly control access to your cloud services and ensure that users have the necessary permissions to perform their tasks.

Implementing these IAM best practices will allow you effectively manage user access to your cloud resources and minimize the risk of unauthorized access. A strong IAM strategy is essential for maintaining a secure and compliant cloud environment, ensuring that your data and resources remain protected.

Threat Detection and Incident Response

In the ever-evolving landscape of cybersecurity, it is crucial to stay vigilant and proactive in monitoring, detecting, and responding to security threats in the cloud. In this section, we'll explore how you can use cloud-native security tools and third-party solutions to enhance your threat detection capabilities, scan for vulnerabilities, and automate your response to potential security incidents.

- **Monitoring and Logging:** Continuous monitoring of your cloud environment is essential for detecting potential security threats. Enable logging for your cloud services, and collect logs in a centralized location for analysis. Cloud providers often offer monitoring and logging services, such as Amazon CloudWatch, Azure Monitor, or Google Cloud Logging, which can help you track and analyze events in real-time.
- **Threat Detection:** Implement threat detection tools to identify suspicious activities or patterns that may indicate a security breach. Many cloud providers offer native threat detection services, such as AWS GuardDuty, Azure Security Center, or Google Cloud Security Command Center. Additionally, consider using third-party solutions like intrusion detection systems (IDS) or security information and event management (SIEM) tools to further enhance your threat detection capabilities.
- **Vulnerability Scanning:** Regularly scan your cloud infrastructure for vulnerabilities that could be exploited by attackers. Use cloud-native or third-party vulnerability scanners to identify misconfigurations, outdated software, or unpatched vulnerabilities in your environment. Address any identified vulnerabilities promptly to reduce the risk of a security breach.
- **Incident Response Plan:** Develop a well-defined incident response plan to guide your actions in the event of a security breach. Your plan should outline the steps to take, roles and responsibilities, communication channels, and escalation procedures. Regularly review and update your incident response plan to ensure it remains effective and relevant.

- **Automated Response:** Leverage cloud-native or third-party tools to automate your response to security incidents. Automating tasks such as isolating affected resources, blocking malicious traffic, or revoking compromised credentials can help you contain security breaches more quickly and minimize the potential damage.

By implementing these strategies, you can strengthen your cloud environment's security posture and be better prepared to detect and respond to security threats. Proactively monitoring, detecting, and responding to potential security incidents is a crucial aspect of maintaining a secure and resilient cloud infrastructure.

Compliance and Governance

Navigating the complex world of regulatory compliance is a critical aspect of managing your cloud environment. Compliance with industry standards such as GDPR, HIPAA, and PCI-DSS ensures that your organization meets the necessary security and privacy requirements for handling sensitive data. In this section, we'll discuss the importance of regulatory compliance, explore cloud provider compliance certifications, and outline best practices for managing security policies, audit logs, and other governance-related tasks.

- **Regulatory Compliance:** Depending on your industry and the type of data you handle, your organization may need to comply with various regulations and standards. These requirements are designed to protect sensitive information and maintain user privacy. Adhering to regulatory compliance helps you avoid fines, legal penalties, and reputational damage that can result from non-compliance.

- **Cloud Provider Compliance Certifications:** When selecting a cloud provider, it's essential to ensure they have the necessary compliance certifications for your industry. Major cloud providers such as AWS, Azure, and Google Cloud Platform typically have a range of certifications, including GDPR, HIPAA, and PCI-DSS, among others. Verify that your provider's certifications meet your organization's requirements before moving your data and workloads to the cloud.

- **Security Policies:** Develop and implement comprehensive security policies that outline the required controls and procedures to maintain compliance with industry standards. These policies should cover areas such as data protection, access control, network security, and incident response. Regularly review and update your policies to keep them relevant and effective.

- **Audit Logs:** Maintain a detailed record of all activities in your cloud environment by enabling audit logging for your cloud services. Audit logs help you track user actions, identify potential security issues, and provide evidence of compliance during audits. Use cloud-native or third-party tools to analyze your logs and detect anomalies that may indicate security breaches or non-compliant behavior.

- **Governance-related Tasks:** Establish processes for managing and maintaining your cloud environment in a compliant manner. This includes tasks such as patch management, vulnerability scanning, and user access reviews. Regularly assess your environment's compliance status and remediate any gaps that may arise.

Understanding the significance of regulatory compliance and following best practices for managing security policies, audit logs, and other governance-related tasks, will allow you to ensure that your cloud environment meets the necessary requirements and maintains a strong security posture. Compliance and governance play a vital role in protecting your sensitive data and building trust with your customers and partners

Implementing Identity and Access Management (IAM)

In this section, we will explore the vital role of Identity and Access Management (IAM) in securing cloud environments and ensuring that only authorized users have access to the appropriate resources. As organizations continue to adopt cloud-based infrastructure and services, the need for a robust and scalable IAM solution becomes increasingly important. Effective IAM implementation is crucial in preventing unauthorized access, data breaches, and maintaining regulatory compliance.

IAM is a comprehensive framework for managing user identities, authentication, authorization, and access control within an organization. By defining and enforcing access policies, IAM ensures that the right users have access to the right resources at the right time. IAM plays a critical role in securing cloud environments, as it provides a centralized way to manage access to various cloud services and resources.

To illustrate the importance of IAM, consider a scenario where an organization has multiple teams working on different projects, each with varying levels of access to cloud resources. Without proper IAM implementation, there is a risk of unauthorized access or accidental exposure of sensitive data. In addition, the organization may struggle to meet compliance requirements, as it would be difficult to track and audit user activity in the cloud environment.

Implementing IAM will allow you create and manage user identities, assign them to specific groups or roles, and define access policies that dictate what resources each user can access. With IAM in place, the organization can ensure that users only have the necessary permissions to perform their tasks, reducing the risk of security breaches and data leaks. Moreover, IAM allows for streamlined auditing and monitoring of user activity, enabling the organization to maintain compliance with industry regulations.

In the following sections, we will delve deeper into the essential components and concepts of IAM, discuss best practices for its implementation, and provide practical examples of how to configure and manage IAM settings in leading cloud platforms. This will equip you with the knowledge and skills needed to implement a secure and effective IAM strategy for your organization's cloud environment.

Importance of IAM in Cloud Security

The importance of Identity and Access Management (IAM) in cloud security cannot be overstated. As you migrate your applications and data to the cloud, effective implementation of IAM is essential for protecting your resources and maintaining compliance with regulatory standards. In this section, we will discuss the critical role that IAM plays in securing your cloud environment and the various ways it helps in controlling user access, ensuring secure authentication, and maintaining compliance.

- **Controlling User Access:** A well-implemented IAM system allows you to have granular control over who can access your cloud resources and what actions they can perform. By defining and managing users, groups, and roles, you can ensure that only authorized individuals have access to the appropriate resources. This minimizes the risk of unauthorized access, data breaches, and security incidents. For example, you can assign specific permissions to a user or a group of users, granting them access to specific cloud resources, such as compute instances or storage buckets, while restricting access to others.
- **Ensuring Secure Authentication:** IAM plays a crucial role in authenticating users before they access your cloud

resources. By implementing strong authentication methods, such as multi-factor authentication (MFA), you can add an additional layer of security to protect your resources. For instance, you can require users to provide not only their username and password but also a one-time passcode generated by an authenticator app or received via text message. This reduces the likelihood of unauthorized access due to compromised credentials.

- **Maintaining Compliance:** Adhering to industry-specific regulations and standards is critical for organizations operating in the cloud. Implementing IAM effectively helps you to maintain compliance by providing audit trails, monitoring user activity, and ensuring that access policies are in line with regulatory requirements. For example, by using IAM to manage access to sensitive data, you can demonstrate compliance with data protection regulations such as GDPR or HIPAA.

In summary, effective implementation of IAM is a cornerstone of cloud security, as it enables you to control user access, ensure secure authentication, and maintain compliance with relevant standards. By understanding the significance of IAM in securing your cloud resources and applying best practices, you can mitigate potential security risks and protect your valuable assets in the cloud.

IAM Components and Concepts

Understanding the core components and concepts of IAM is essential for implementing proper access control and secure resource management in your cloud environment. In this section, we will introduce you to the fundamental elements of IAM, including users, groups, roles, permissions, policies, and multi-factor authentication. We will also describe how these components work together to ensure a robust and effective IAM solution.

- **Users:** Users represent individual accounts in the IAM system. Each user has a unique identifier and a set of credentials, such as a username and password, for authentication purposes. When configuring IAM, you will create and manage user accounts for everyone who needs access to your cloud resources, assigning appropriate permissions based on their job responsibilities.

- **Groups:** Groups are a way to organize users with similar access requirements. By adding users to a group, you can assign and manage permissions collectively, simplifying the administration of access policies. For example, you might create a group for your development team and grant them access to specific resources required for their tasks.

- **Roles:** Roles are another method for managing access permissions. Instead of assigning permissions directly to users, you can define a role with a specific set of permissions and then grant the role to users or groups. This approach promotes the principle of least privilege, ensuring that users only have the necessary permissions to perform their tasks. Roles can also be used to grant permissions to applications or services within your cloud environment.

- **Permissions:** Permissions define the actions that a user, group, or role is allowed to perform on a specific resource. In IAM, you will define permissions for various cloud resources, such as compute instances, storage buckets, or databases. By carefully assigning permissions, you can maintain fine-grained control over who can access and interact with your cloud resources.

- **Policies:** Policies are documents that formalize the permissions for users, groups, or roles. They define which actions are allowed or denied for specific resources. When you create a policy, you will specify the permissions that are granted to the associated users, groups, or roles, effectively defining their access rights in the cloud environment.

- **Multi-Factor Authentication (MFA):** MFA is an additional layer of security that requires users to provide more than one form of identification to authenticate their access. By enabling MFA, you can significantly enhance the security of your cloud resources, as it reduces the risk of unauthorized access due to compromised credentials.

Together, these IAM components and concepts provide a comprehensive framework for managing access to your cloud resources. By understanding how these elements work in conjunction, you can create a secure and efficient IAM system that ensures proper access control and resource management in your cloud environment.

IAM Best Practices

Implementing IAM effectively requires adhering to best practices that promote security and maintain compliance. In this section, we will discuss some of the most important best practices for implementing IAM in your cloud environment, including the principle of least privilege, regular audits of access rights, the use of strong authentication methods, and implementing separation of duties. We will also emphasize the importance of continuous monitoring and improvement of your IAM policies.

- **Principle of Least Privilege:** This fundamental security principle states that you should grant users, groups, or roles only the minimum set of permissions necessary to perform their tasks. By following this principle, you can minimize the potential damage caused by unauthorized access or compromised credentials. When assigning permissions, start by granting the most restrictive access and only add additional permissions as needed for specific tasks.

- **Regular Audits of Access Rights:** Conducting regular audits of your IAM policies and access rights ensures that your access controls remain up-to-date and secure. By periodically reviewing your user accounts, groups, roles, and permissions, you can identify and address any inconsistencies or unnecessary access rights. This practice also helps you maintain compliance with regulatory standards that require ongoing monitoring of access controls.

- **Strong Authentication Methods:** Implementing strong authentication methods, such as multi-factor authentication (MFA), is a critical IAM best practice. By requiring users to provide multiple forms of identification, you can significantly enhance the security of your cloud resources. Encourage the use of strong, unique passwords and implement MFA to provide an additional layer of protection against unauthorized access.

- **Separation of Duties:** This best practice involves dividing critical tasks and responsibilities among multiple users or roles to prevent any single individual from having too much control or access to sensitive resources. By implementing separation of duties, you can reduce the risk of fraud, errors, and security breaches. For example, you might separate the responsibilities of deploying applications and managing database access, ensuring that no single person has complete control over both tasks.

- **Continuous Monitoring and Improvement:** IAM is an ongoing process that requires continuous monitoring and improvement. Regularly review your IAM policies, access controls, and user activity to identify potential security risks and areas for improvement. Stay informed about the latest security threats and IAM best practices to ensure that your cloud environment remains secure and compliant.

Following these IAM best practices will allow you create a secure and efficient access control system that protects your cloud resources and maintains regulatory compliance. By understanding the importance of these practices and implementing them in your cloud environment, you will be well-equipped to manage access and mitigate potential security risks.

IAM Implementation in Major Cloud Providers

Implementing IAM in your cloud environment requires understanding the IAM services offered by major cloud providers, such as AWS, Azure, and Google Cloud Platform. In this section, we will provide an overview of the IAM services provided by these leading platforms, highlight the similarities and differences in their approach to IAM, and discuss how you can configure IAM settings in each platform.

- **AWS Identity and Access Management (IAM):** AWS offers a comprehensive IAM service that allows you to create and manage users, groups, roles, and policies. You can define fine-grained permissions for various AWS resources and services. To configure IAM in AWS, you can use the AWS Management Console, AWS CLI, or SDKs. Some key features of AWS IAM include:

    - Role-based access control (RBAC)
    - Integration with AWS Organizations for centralized management
    - Support for multi-factor authentication (MFA)
    - Integration with AWS CloudTrail for auditing and monitoring

- **Azure Active Directory (Azure AD):** Azure AD is Microsoft's cloud-based identity and access management service. It offers a wide range of features for managing user identities, access control, and authentication in Azure and other Microsoft services. You can configure Azure AD using the Azure Portal, Azure CLI, or SDKs. Some key features of Azure AD include:

- Role-based access control (RBAC)
- Support for multi-factor authentication (MFA)
- Integration with on-premises Active Directory for hybrid environments
- Conditional Access policies for adaptive access control

- **Google Cloud Identity and Access Management (Cloud IAM):** Cloud IAM is Google Cloud Platform's IAM service, enabling you to manage access control for your Google Cloud resources. You can create and manage users, groups, roles, and policies in the Google Cloud Console, gcloud CLI, or API. Some key features of Google Cloud IAM include:

    - Role-based access control (RBAC)
    - Support for custom roles and permissions
    - Integration with Google Workspace for user and group management
    - Audit logging with Google Cloud's operations suite

While each cloud provider has its unique approach to IAM, there are similarities across these platforms, such as role-based access control, support for multi-factor authentication, and integration with various auditing and monitoring tools. The primary differences lie in their specific implementation details, user interfaces, and integration with other services offered by each provider.

To configure IAM settings in each platform, you should familiarize yourself with their respective management consoles, CLI tools, and SDKs. By understanding the IAM services offered by these leading cloud providers, you can implement a robust and secure IAM solution tailored to your cloud environment and specific requirements.

IAM Integration with Third-Party Services

Integrating IAM with third-party services is essential for organizations that want to centralize user management and simplify access control across multiple cloud services and applications. In this section, we will explain how IAM can be integrated with external identity providers, such as LDAP or Active Directory, and discuss the role of federated identities and Single Sign-On (SSO) in streamlining access management.

- **Integration with External Identity Providers:** Many organizations use directory services like LDAP or Active Directory to manage user identities and access controls on-premises. You can integrate your cloud IAM solution with these external identity providers to centralize user management and maintain a single source of truth for user information. This integration allows you to leverage existing user accounts and access controls, reducing the administrative overhead of managing users in multiple systems. Major cloud providers, such as AWS, Azure, and Google Cloud, offer built-in integration options with popular directory services.

- **Federated Identities:** Federated identity is a method for linking user identities across multiple systems and organizations. By implementing federated identity management, you can share authentication and authorization information between your on-premises directory service and your cloud IAM solution. This allows users to authenticate using their existing credentials and access cloud resources based on their assigned permissions in the on-premises directory. Federated identities help you maintain consistent access control policies across your entire IT environment, both on-premises and in the cloud.

- **Single Sign-On (SSO):** Single Sign-On is a feature that allows users to authenticate once and gain access to multiple applications and services without needing to re-enter their credentials. By implementing SSO, you can simplify the user experience and reduce the burden of remembering multiple passwords. SSO can be used across different cloud providers and third-party applications, enabling a seamless and secure experience for users as they navigate between various services. Popular SSO protocols include SAML, OAuth, and OpenID Connect, which are supported by most cloud IAM solutions and external identity providers.

For example, you can configure AWS, Azure, or Google Cloud IAM to use your on-premises Active Directory as the identity provider. Users can then authenticate using their existing Active Directory credentials and access cloud resources based on the permissions granted to them. With SSO enabled, users can seamlessly navigate between cloud services and on-premises applications without having to re-authenticate each time.

In summary, integrating IAM with third-party services like LDAP or Active Directory and leveraging federated identities and Single Sign-On can greatly simplify access management across multiple cloud services and applications. By understanding these concepts and incorporating them into your IAM strategy, you can create a secure and user-friendly access control system that streamlines the management of your cloud resources.

Practical: Configuring Security Policies and Monitoring

In this practical section we will dive into the hands-on aspects of implementing Identity and Access Management (IAM) in your cloud environment. By mastering these skills, you will be better equipped to secure your cloud resources, maintain regulatory compliance, and manage access to services effectively. We will cover essential aspects of IAM configuration, such as creating and managing policies, setting up multi-factor authentication, configuring access logging and auditing, and implementing monitoring and alerts for IAM-related events.

Throughout this section, we will provide practical examples and step-by-step guidance to help you apply these concepts to your specific cloud environment. By the end of this section, you will have gained valuable hands-on experience and a deeper understanding of IAM best practices, enabling you to create a robust and secure access control system tailored to your organization's needs. With this knowledge, you can confidently manage your cloud resources while mitigating potential security risks and maintaining compliance with industry standards.

Creating and Managing IAM Policies

Creating and managing IAM policies is a crucial aspect of controlling access to your cloud resources and services. In this section, we will guide you through the process of creating, modifying, and applying IAM policies, defining permissions, associating policies with users, groups, or roles, and testing the effectiveness of your access controls.

To create an IAM policy, you first need to define the permissions that dictate what actions a user, group, or role can perform on specific resources. Permissions are typically defined using JSON (JavaScript Object Notation) and follow a standard format that includes the action, resource, and optional conditions. For example, you may define a policy that allows read-only access to a specific storage bucket in your cloud environment.

Once you have defined the permissions in a policy, you can associate it with users, groups, or roles within your IAM system. Users are individual accounts with specific credentials, while groups are collections of users that share the same set of permissions. Roles, on the other hand, represent a set of permissions that can be assumed by different entities, such as users or applications, allowing for more flexible and dynamic access control.

To apply an IAM policy, you simply attach it to the desired user, group, or role. This can be done through the management console, CLI, or SDKs provided by your cloud platform. Keep in mind that it's often best to follow the principle of least privilege, granting only the necessary permissions to accomplish a specific task.

Testing the effectiveness of your access controls is an important step in ensuring the security of your cloud environment. You can do this by attempting to perform actions that should be allowed or denied based on your IAM policies. For example, you might log in as a user with read-only access to a storage bucket and verify that you can view the contents but cannot modify or delete any files.

In summary, by learning how to create, modify, and apply IAM policies, you can effectively control access to your cloud resources and services. Understanding the process of defining permissions, associating policies with users, groups, or roles, and testing the effectiveness of your access controls will empower you to create a robust and secure IAM system tailored to your organization's needs.

Configuring Multi-Factor Authentication (MFA)

Configuring Multi-Factor Authentication (MFA) is an important step in enhancing the security of your cloud resources. MFA adds an extra layer of protection by requiring users to provide two or more forms of identification when accessing cloud services. In this section, we will explore the steps to enable and configure MFA for your IAM users, understand the different MFA methods available, and learn how to enforce MFA requirements for specific users or groups.

To set up MFA, you need to choose an MFA method that suits your organization's needs. There are several options available, including:

- **Hardware tokens:** Physical devices that generate one-time passwords (OTPs) to be used alongside user credentials. Hardware tokens are considered highly secure but can be more expensive and cumbersome to manage.
- **Virtual MFA devices:** Software-based solutions that generate OTPs on a user's device, such as a smartphone or a desktop application. Virtual MFA devices offer a balance between security and convenience, as they do not require additional physical hardware.
- **Smartphone apps:** Many MFA apps are available for smartphones, such as Google Authenticator, Authy, or Microsoft Authenticator. These apps generate OTPs on the user's phone, which must be entered during the authentication process.

Once you've chosen an MFA method, follow the steps provided by your cloud platform to enable and configure MFA for your IAM users. This typically involves activating MFA in your IAM settings and then guiding users through the process of registering their MFA devices or apps.

To enforce MFA requirements for specific users or groups, you can create and apply IAM policies that require MFA authentication for certain actions or access to specific resources. For example, you might create a policy that requires MFA for any action that modifies sensitive data or critical infrastructure.

Here's a practical example: If you're using AWS, you can enable MFA for a user by navigating to the IAM console, selecting the user, and then choosing the "Security Credentials" tab. From there, you can click on "Manage" next to "Assigned MFA device" and follow the prompts to associate a hardware token, virtual MFA device, or smartphone app with the user account. You can then create and apply an IAM policy that requires MFA for specific actions or resources.

In conclusion, configuring MFA for your IAM users is an effective way to enhance the security of your cloud resources. By understanding the different MFA methods available and learning how to enforce MFA requirements for specific users or groups, you can implement a robust and secure access control system that helps protect your organization from unauthorized access and potential security breaches.

Setting Up Access Logging and Auditing

Setting up access logging and auditing in your cloud environment is crucial for tracking user activity, identifying potential security risks, and maintaining regulatory compliance. In this section, we will guide you through configuring access logging and auditing and introduce you to the tools and services available for logging and auditing in major cloud platforms.

To set up access logging, you'll need to enable and configure the appropriate logging services provided by your cloud platform. These services record user activity within your environment, creating an audit trail that can be analyzed for security and compliance purposes.

Here are some logging and auditing services offered by major cloud platforms:

- **AWS CloudTrail:** CloudTrail captures AWS API calls made by or on behalf of your AWS account, recording details such as the identity of the caller, the time of the call, and the requested actions. You can configure CloudTrail to deliver log files to an Amazon S3 bucket for storage and analysis.

- **Azure Monitor:** Azure Monitor collects and analyzes log data from Azure resources, enabling you to track user activity, diagnose issues, and gain insights into your environment. You can configure Azure Monitor to send logs to a storage account, stream them to an event hub, or send them to Azure Monitor logs for further analysis.

- **Google Cloud's operations suite:** Google Cloud provides a suite of logging and monitoring tools, including Cloud Logging and Cloud Monitoring, that work together to collect, store, and analyze log data from your Google Cloud resources. You can configure Cloud Logging to export log data to Cloud Storage, BigQuery, or Pub/Sub for further processing and analysis.

To set up access logging and auditing in your environment, follow the steps provided by your cloud platform for enabling and configuring the appropriate services. For example, to enable AWS CloudTrail, navigate to the CloudTrail console, create a new trail, and specify the S3 bucket where you want to store your log files. You can also configure additional settings, such as log file encryption and data event logging.

Once you have configured access logging and auditing, it's essential to regularly review and analyze the log data to identify potential security risks, maintain regulatory compliance, and optimize your environment. This may involve using built-in tools provided by your cloud platform, third-party log analysis tools, or custom solutions tailored to your organization's needs.

In summary, setting up access logging and auditing in your cloud environment is an essential aspect of maintaining security and compliance. By learning how to configure access logging and auditing and familiarizing yourself with the tools and services available in major cloud platforms, you can effectively track user activity, identify potential security risks, and ensure your environment adheres to regulatory requirements.

Implementing IAM Monitoring and Alerts

Implementing IAM monitoring and alerts is a vital aspect of maintaining the security of your cloud environment, as it enables you to detect and respond to potential security incidents quickly. In this section, we will guide you through setting up monitoring and alerts for your IAM policies and access controls, using cloud monitoring tools such as Amazon CloudWatch, Azure Monitor, or Google Cloud Monitoring to create custom IAM-related metrics, dashboards, and alerting rules.

To start setting up IAM monitoring and alerts, you'll need to choose a cloud monitoring tool that suits your needs. As mentioned earlier, some popular options include Amazon CloudWatch, Azure Monitor, and Google Cloud Monitoring. These tools provide robust monitoring capabilities and can be easily integrated with your IAM policies and access controls.

Once you have chosen a monitoring tool, follow the steps provided by your cloud platform to configure IAM-related metrics. This may involve setting up metric filters to collect specific IAM events, such as policy changes, user logins, or permission denials. For example, you might create a metric filter in Amazon CloudWatch to track the number of denied API requests, which could indicate unauthorized attempts to access resources.

Next, create dashboards to visualize your IAM-related metrics, making it easier to monitor the state of your access controls and identify potential issues. Dashboards can be customized to display the most relevant information for your organization, such as the number of active users, policy changes, or access attempts to sensitive resources.

Finally, configure alerting rules to notify you or your team when specific IAM events occur or when metrics cross predefined thresholds. For example, you might set up an alert to notify you when a user modifies a sensitive IAM policy or when there's a sudden spike in denied access requests. Alerts can be sent through various channels, such as email, SMS, or third-party messaging services, depending on your preferences and needs.

Here's a practical example: To set up an alert in Amazon CloudWatch for denied API requests, you can create an alarm based on the custom metric filter you created earlier. Configure the alarm to trigger when the number of denied requests exceeds a certain threshold within a specified time frame, and specify the notification method (e.g., email or SMS) to receive the alert.

So, setting up IAM monitoring and alerts is crucial for detecting and responding to potential security incidents in your cloud environment. By learning how to use cloud monitoring tools to create custom IAM-related metrics, dashboards, and alerting rules, you can effectively safeguard your resources and maintain a secure and compliant cloud infrastructure.

Performing Regular IAM Reviews and Updates

Performing regular IAM reviews and updates is a critical aspect of maintaining a secure and compliant cloud environment. As your organization evolves, it's essential to ensure that your IAM policies, user accounts, and access rights remain up-to-date and secure. In this section, we will discuss the importance of conducting periodic IAM reviews, best practices for performing IAM audits, and how to update your IAM configuration to maintain compliance and mitigate security risks.

Periodic IAM reviews are crucial for several reasons:

- **Identifying and addressing inconsistencies:** Regular audits help you spot potential inconsistencies in your IAM policies, such as unintended permissions or orphaned accounts. Addressing these inconsistencies can reduce security risks and ensure that your access controls align with your organization's needs.
- **Maintaining compliance:** Many regulatory frameworks require organizations to periodically review and update their access controls. Regular IAM audits help you stay compliant with these requirements and demonstrate your commitment to security.
- **Adapting to changes in your organization:** As your organization grows and changes, your IAM configuration should evolve as well. Regular reviews help you identify areas where access rights need to be updated to reflect new roles, responsibilities, or business requirements.

To perform an effective IAM review, follow these best practices:

1. **Establish a review schedule:** Determine how often you will conduct IAM reviews, based on your organization's needs and regulatory requirements. This could be monthly, quarterly, or annually.
2. **Review user accounts and access rights:** Examine each user account to ensure it is still needed and that the associated access rights are appropriate for the user's role. Deactivate or delete accounts that are no longer required, and update access rights as necessary.
3. **Check for inconsistencies:** Look for potential issues such as overly permissive policies, unused permissions, or users with access to resources they should not have.

Address these issues by adjusting policies or access rights as needed.

4. **Validate changes:** After making updates to your IAM configuration, test the changes to ensure they work as intended and do not introduce new security risks.

5. **Document your findings:** Keep a record of the results of your IAM audits, including any issues identified and the actions taken to address them. This documentation can be useful for demonstrating compliance and providing a historical context for future audits.

Here's a practical example: As part of your regular IAM review, you might discover that a user who has changed roles within your organization still has access to resources that are no longer relevant to their current position. In this case, you would update their access rights to align with their new role and remove any unnecessary permissions.

In conclusion, performing regular IAM reviews and updates is essential for maintaining a secure and compliant cloud environment. By understanding the importance of periodic reviews, following best practices for IAM audits, and updating your IAM configuration as needed, you can effectively manage access to your cloud resources and minimize security risks.

Module 4 Exam Questions

1. What is the primary goal of Identity and Access Management (IAM)?

a) Resource monitoring

b) Secure authentication and access control

c) Cost optimization

d) Application deployment

2. Which component of IAM involves assigning permissions to a user, group, or role?

a) Policies

b) Roles

c) Users

d) Groups

3. Which principle states that a user should only have the minimum level of access required to perform their job?

a) Separation of duties

b) Principle of least privilege

c) Access management

d) Role-based access control

4. Which IAM feature requires users to provide two or more separate forms of identification during the authentication process?

a) Multi-factor authentication (MFA)

b) Single Sign-On (SSO)

c) Federated identity

d) Access control

5. AWS IAM, Azure Active Directory, and Google Cloud IAM are examples of what?

a) Containerization platforms

b) Serverless computing services

c) IAM services in major cloud providers

d) Database management systems

6. Which of the following is a best practice for implementing IAM?

a) Granting all users full access to resources

b) Sharing user accounts among team members

c) Regularly auditing access rights

d) Avoiding the use of MFA

7. LDAP and Active Directory are examples of what?

a) External identity providers

b) Monitoring tools

c) CI/CD tools

d) Cloud providers

8. Amazon CloudWatch, Azure Monitor, and Google Cloud Monitoring are examples of what?

a) IAM services

b) Cloud monitoring tools

c) Database management systems

d) Serverless computing services

9. What is the purpose of IAM access logging and auditing?

a) To track user activity and maintain compliance

b) To optimize cloud costs

c) To deploy applications

d) To manage databases

10. Which cloud service is used for logging and auditing in AWS?

a) AWS CloudTrail

b) Amazon CloudWatch

c) AWS IAM

d) AWS Lambda

11. Which cloud service is used for logging and auditing in Azure?

a) Azure Active Directory

b) Azure Functions

c) Azure Monitor

d) Azure Kubernetes Service

12. Which cloud service is used for logging and auditing in Google Cloud Platform?

a) Google Cloud Identity and Access Management

b) Google Cloud Functions

c) Google Cloud's operations suite

d) Google Kubernetes Engine

13. Which of the following methods can be used for multi-factor authentication (MFA)?

a) Password

b) Hardware token

c) Virtual MFA device

d) All of the above

14. What is the main purpose of Single Sign-On (SSO)?

a) To simplify access management across multiple services

b) To enforce MFA requirements

c) To track user activity

d) To optimize cloud costs

15. Which IAM component is used to group users with similar access requirements?

a) Roles

b) Groups

c) Policies

d) Permissions

16. Which of the following should be included in a regular IAM review?

a) Reviewing user accounts and access rights

b) Checking for inconsistencies

c) Validating changes

d) All of the above

17. What is the purpose of creating dashboards for IAM-related metrics?

a) To visualize access controls and identify potential issues

b) To deploy applications

c) To manage databases

d) To optimize cloud costs

18. Which of the following is a benefit of integrating IAM with third-party services like LDAP or Active Directory?

a) Centralized user management

b) Reduced security

c) Higher cloud costs

d) Slower application deployment

19. Which IAM component defines the actions a user, group, or role can perform on specific resources?

a) Policies

b) Users

c) Groups

d) Roles

20. What is the main objective of IAM monitoring and alerts?

a) To detect and respond to potential security incidents

b) To manage application deployment

c) To optimize cloud costs

d) To improve database performance

Answers:

1. b
2. a
3. b
4. a
5. c
6. c
7. a
8. b
9. a
10. a
11. c
12. c
13. d
14. a
15. b
16. d
17. a
18. a
19. a
20. a

Module 5. Deploying and Managing Applications

Introduction to Deploying and Managing Applications

In this module, we will explore the essential skills required to successfully deploy, manage, and scale applications in the cloud. With the increasing adoption of cloud technologies, it's crucial for IT professionals, developers, and system administrators to have a deep understanding of how to efficiently deploy and manage applications in a cloud environment.

In this module, we will discuss the concept of containerization and how it has revolutionized the way applications are built, packaged, and deployed. Containerization ensures that applications run consistently across different environments, thereby reducing the complexities of deployment and maintenance. We will explore popular containerization platforms such as Docker and dive into the world of container orchestration with Kubernetes.

Next, we will delve into the realm of serverless computing, a game-changing approach that allows developers to focus on their code without worrying about the underlying infrastructure. We will examine popular serverless platforms such as AWS Lambda, Azure Functions, and Google Cloud Functions, and discuss their benefits, use cases, and limitations.

Throughout this module, we will provide practical examples to illustrate the concepts discussed. You will learn how to deploy a sample application using containerization and orchestration tools, as well as how to build and deploy a serverless application using popular cloud platforms.

By the end of this module, you will have gained a solid understanding of the key technologies and methodologies used to deploy and manage applications in the cloud. You will be equipped with the knowledge and hands-on experience to confidently navigate the ever-evolving landscape of cloud application deployment and management.

Containerization and Orchestration (Docker, Kubernetes)

Containerization has emerged as a powerful solution to address the challenges associated with deploying and managing applications consistently across various environments. In this section, we will introduce the concept of containerization, discuss its benefits, and explore how it has transformed the way applications are developed and managed.

Containerization is the process of encapsulating an application, along with its dependencies, libraries, and runtime environment, into a single, lightweight, and portable unit called a container. Containers run in isolation from one another, sharing only the underlying host operating system's resources, which makes them highly efficient, scalable, and easy to manage.

The rise of containerization can be attributed to several key benefits:

- **Consistent execution:** Containers enable applications to run consistently across different environments, such as development, testing, and production, by packaging the application code and its dependencies together. This eliminates the "it works on my machine" problem and streamlines the deployment process.
- **Isolation:** Containers run in isolation from each other and the host system, ensuring that application conflicts, such as dependency version clashes, are minimized. This isolation also enhances security by limiting the potential attack surface.
- Portability: Since containers encapsulate the application and its environment, they can be easily transferred between different host systems and cloud platforms, promoting flexibility and reducing vendor lock-in.

- **Resource efficiency:** Containers share the host operating system's resources, such as the kernel, which results in lower resource overhead compared to traditional virtualization techniques. This makes containers ideal for deploying applications at scale, without incurring excessive costs.

A practical example of containerization is the deployment of a web application that relies on a specific version of a web server, a database, and a set of libraries. By packaging the application and its dependencies into a container, developers can ensure that the application will run consistently across different environments, regardless of the local setup and configurations.

In conclusion, containerization has revolutionized the way applications are built, packaged, and deployed, offering increased consistency, isolation, portability, and resource efficiency. In the following sections, we will delve deeper into the world of containerization and explore popular tools, such as Docker and Kubernetes, that have become instrumental in managing and orchestrating containerized applications.

Docker basics

In this section, we will introduce you to Docker, a leading containerization platform that has become synonymous with the container revolution. Docker has made it easier for developers, system administrators, and IT professionals to embrace containerization and leverage its benefits for application deployment and management. We will discuss the fundamentals of Docker, including its architecture, components, and key commands, and demonstrate how to create and manage Docker images, containers, and Dockerfiles.

Docker is an open-source platform that automates the process of creating, deploying, and running applications inside lightweight, portable containers. Docker's architecture is built on a client-server model, where the Docker client communicates with the

Docker daemon, which, in turn, manages the creation, management, and execution of containers. The Docker client and daemon can run on the same host, or they can communicate over a network.

Some of the key components of Docker include:

- **Docker images:** These are read-only templates used to create containers. Images are built from a set of instructions specified in a Dockerfile and can be stored in a registry, such as Docker Hub or a private registry, for easy distribution and reuse.

- **Docker containers:** These are the running instances of Docker images. Containers are isolated, portable units that package an application and its dependencies, providing a consistent runtime environment across different platforms.

- **Dockerfiles:** These are text files containing a series of instructions used to create a Docker image. Dockerfiles specify the base image, application code, dependencies, and configuration details required to build a customized image for an application.

To illustrate the practical usage of Docker, consider a scenario where a developer wants to deploy a Python web application. The developer can create a Dockerfile specifying the base Python image, copy the application code into the image, and install any required dependencies. Next, the developer can use Docker commands to build an image from the Dockerfile and run a container from the image, ensuring that the application executes consistently across different environments.

In summary, this section will provide you with a solid foundation in Docker basics, covering its architecture, components, and key commands. By understanding how to create and manage Docker images, containers, and Dockerfiles, you will be better equipped to leverage the power of containerization to simplify application deployment and management. In the subsequent sections, we will further explore the world of containerization, focusing on container orchestration with Kubernetes.

as you gain proficiency in Docker and understand the core concepts, you'll inevitably encounter scenarios where you need to manage multiple containers that interact with each other. Manually managing a large number of containers can be a challenging task, especially when it comes to scaling, updates, and fault tolerance. This is where container orchestration comes into play.

Container orchestration is the automated process of deploying, scaling, and managing containerized applications. It involves coordinating the lifecycle of containers, ensuring that they are running as intended, and automating tasks such as scaling, load balancing, and updates. A popular container orchestration platform is Kubernetes, which has become the de facto standard for managing containerized applications at scale.

In the upcoming sections, we will dive into Kubernetes and discuss its architecture, components, and key features. You will learn how to create and manage Kubernetes resources such as pods, services, and deployments, and explore advanced topics like autoscaling and rolling updates. By understanding the capabilities of Kubernetes and how it integrates with Docker, you will be able to harness the full potential of containerization and orchestration to deploy, manage, and scale applications efficiently and effectively.

As you progress through the module, remember to engage with the practical examples provided. These hands-on exercises will solidify your understanding of the concepts and help you develop the skills necessary to confidently work with containerization and orchestration tools like Docker and Kubernetes in real-world scenarios.

Docker in AWS, Microsoft Azure, and Google Cloud

Docker is widely used across major cloud providers, such as AWS, Microsoft Azure, and Google Cloud, to build, package, and deploy containerized applications. Each cloud provider offers services and tools that integrate with Docker, allowing developers to manage containers efficiently and effectively. In this section, we will discuss how Docker is used in each of these cloud providers, providing practical examples and configuration details.

AWS:

Amazon Web Services (AWS) offers several services that support Docker container management, including Amazon Elastic Container Service (ECS), AWS Fargate, and Amazon Elastic Kubernetes Service (EKS).

- **Amazon ECS:** ECS is a fully-managed container orchestration service that supports Docker containers. You can create task definitions in JSON format, which specify the Docker images, resource requirements, and other configurations for your containers. Using Amazon ECS, you can run, scale, and monitor your Docker containers across a fleet of Amazon EC2 instances.

  Practical Example: To run a Docker container with Amazon ECS, follow these steps:

  - Create a task definition in JSON format, specifying the Docker image, memory and CPU requirements, and other configurations.
  - Create an ECS cluster using the AWS Management Console or AWS CLI.
  - Launch and manage your containers using the task definition and the created ECS cluster.

- **AWS Fargate:** Fargate is a serverless compute engine for containers, which allows you to run Docker containers without managing the underlying infrastructure. With Fargate, you can focus on defining your application and its requirements, while AWS handles the provisioning, scaling, and management of the container infrastructure.

Practical Example: To deploy a Docker container with AWS Fargate, follow these steps:

- Create a task definition, similar to the one used with Amazon ECS.
- Choose the Fargate launch type when creating your ECS service.
- Configure the required settings, such as VPC and security groups.

Microsoft Azure:

Microsoft Azure provides Azure Container Instances (ACI) and Azure Kubernetes Service (AKS) to manage Docker containers.

- **Azure Container Instances (ACI):** ACI allows you to run Docker containers without the need for managing virtual machines or container orchestrators. It offers a simple and fast way to deploy containers in Azure with minimal configuration.

  Practical Example: To deploy a Docker container with Azure Container Instances, follow these steps:

  - Use the Azure CLI to create a new container instance, specifying the Docker image, resource requirements, and other configurations.
  - Monitor and manage your container instance using the Azure Portal or Azure CLI.

- **Azure Kubernetes Service (AKS):** AKS is a managed Kubernetes service that supports Docker containers. It simplifies Kubernetes cluster deployment and management, allowing you to focus on your containerized applications.

Practical Example: To deploy a Docker container with Azure Kubernetes Service, follow these steps:

- Create a Kubernetes manifest file, specifying the Docker image and other configurations for your application.
- Use the Azure CLI or Kubernetes CLI (kubectl) to create and manage the resources defined in the manifest file.

Google Cloud:

Google Cloud offers services such as Google Kubernetes Engine (GKE) and Cloud Run to manage Docker containers.

- **Google Kubernetes Engine (GKE):** GKE is a managed Kubernetes service that supports Docker containers. It simplifies Kubernetes cluster deployment and management, allowing you to focus on your containerized applications.

  Practical Example: To deploy a Docker container with Google Kubernetes Engine, follow these steps:

  - Create a Kubernetes manifest file, specifying the Docker image and other configurations for your application.
  - Use the Google Cloud CLI (gcloud) or Kubernetes CLI (kubectl) to create and manage the resources defined in the manifest file.

- **Cloud Run:** Cloud Run is a fully managed serverless platform that allows you to run Docker containers without worrying about the underlying infrastructure. It automatically scales your containers based on incoming traffic and supports both HTTP and gRPC services.

Practical Example: To deploy a Docker container with Google Cloud Run, follow these steps:

- Build and push your Docker image to Google Container Registry (GCR) or another container registry.
- Use the Google Cloud CLI (gcloud) to deploy the container to Cloud Run, specifying the image and other configurations.
- c. Monitor and manage your container instances using the Google Cloud Console or the gcloud CLI.

In conclusion, Docker is widely supported across major cloud providers, such as AWS, Microsoft Azure, and Google Cloud. Each provider offers services and tools that integrate with Docker, allowing you to build, package, and deploy containerized applications seamlessly. By understanding the different services and their usage, you can choose the most suitable approach for deploying and managing your Docker containers in the cloud.

Kubernetes overview

In this section, we will introduce you to Kubernetes, a robust and widely-adopted container orchestration platform that has transformed the way containerized applications are managed at scale. Kubernetes automates the deployment, scaling, and management of containerized applications, helping organizations achieve greater efficiency, reliability, and flexibility in their infrastructure. We will discuss the core components of Kubernetes, such as nodes, pods, services, and deployments, and explain their roles in a Kubernetes cluster.

Kubernetes was originally developed by Google and is now maintained by the Cloud Native Computing Foundation (CNCF). It provides a set of abstractions and primitives that simplify the management of containerized applications, allowing developers and operators to focus on their applications rather than the underlying infrastructure.

Some of the key components of a Kubernetes cluster include:

- **Nodes:** These are the worker machines that run containers. Nodes can be physical or virtual machines, and they provide the necessary resources, such as CPU, memory, and storage, required to execute containerized applications.
- **Pods:** The smallest and simplest unit in the Kubernetes object model, pods represent a single instance of a running application. A pod can contain one or more containers that share the same network namespace and storage volumes, allowing them to communicate and share data efficiently.
- **Services:** These are Kubernetes objects that provide stable network endpoints for pods, enabling communication between different components of an application, as well as with external clients. Services abstract the underlying pods' network details, making it easy to scale or update the application without impacting its consumers.
- **Deployments:** These are high-level Kubernetes objects that describe the desired state of an application, such as the number of replicas, the container image to use, and update strategies. Deployments ensure that the application's actual state matches the desired state by creating, updating, or deleting pods as needed.

To illustrate the practical usage of Kubernetes, consider a scenario where an organization wants to deploy a containerized web application that consists of a front-end, a back-end API, and a database. Using Kubernetes, they can create deployments to manage the desired state of the front-end and back-end components, and services to enable communication between them, as well as with external clients. As the application scales, Kubernetes automatically handles the creation, distribution, and management of additional pods across the available nodes, ensuring optimal resource utilization and fault tolerance.

In conclusion, this section will provide you with a comprehensive understanding of Kubernetes as a powerful container orchestration platform. By exploring its core components and their roles in a Kubernetes cluster, you will be better equipped to manage containerized applications at scale, harnessing the full potential of containerization and orchestration. In the following sections, we will dive deeper into Kubernetes and explore more advanced topics, such as deploying applications, configuring storage, and implementing rolling updates.

Deploying applications with Kubernetes

In this section, we will explore the process of deploying applications using Kubernetes. As a powerful container orchestration platform, Kubernetes provides a flexible and efficient way to manage the lifecycle of containerized applications. We will discuss how to define and deploy applications using Kubernetes manifests and dive into the creation of deployments, services, and ingress resources to manage application accessibility and load balancing.

Kubernetes manifests are YAML or JSON files that describe the desired state of Kubernetes resources, such as deployments, services, and ingress. These manifests provide a declarative approach to application management, allowing developers and operators to specify the desired configuration and behavior of their applications while relying on Kubernetes to ensure the actual state matches the desired state.

A typical application deployment with Kubernetes involves the following steps:

- **Define a deployment manifest:** Create a YAML file that describes the desired state of the application, including the container image, the number of replicas, and any configuration data or environment variables. The deployment manifest instructs Kubernetes on how to create and manage the necessary pods to run the application.

- **Create a service manifest:** Define a YAML file that specifies the type of service (e.g., ClusterIP, NodePort, or LoadBalancer) and any required configuration, such as ports and selectors. The service manifest ensures that the application components can communicate with each other and that external clients can access the application as needed.

- **Configure ingress resources:** For applications that need to be exposed externally, create an ingress manifest that defines the rules for routing external traffic to the appropriate services within the cluster. Ingress resources provide load balancing, SSL termination, and name-based virtual hosting.

To illustrate a practical example, consider a scenario where a developer wants to deploy a web application with a front-end and a back-end API. The developer can create a deployment manifest for each component, specifying the container image, the number of replicas, and any required configuration data. Next, the developer can define a service manifest for each component to enable communication between the front-end and the back-end API, as well as to expose the front-end to external clients. Finally, the developer can create an ingress resource to route external traffic to the front-end service, providing load balancing and SSL termination.

In conclusion, this section will equip you with the knowledge and skills required to define and deploy applications using Kubernetes manifests. By understanding how to create deployments, services, and ingress resources, you will be able to efficiently manage the accessibility and load balancing of your containerized applications, leveraging the full potential of Kubernetes as a powerful container orchestration platform. In the following sections, we will explore more advanced Kubernetes concepts and best practices, such as configuring persistent storage, implementing rolling updates, and setting up autoscaling.

Advanced Kubernetes concepts and best practices

In this section, we will dive deeper into the world of Kubernetes and explore advanced concepts and best practices that can enhance the management, scalability, and reliability of your containerized applications. By understanding these advanced topics, you will be better equipped to optimize application performance and resource utilization, as well as to ensure the smooth operation of your applications in a Kubernetes cluster.

Some of the advanced topics and best practices we will cover in this section include:

- **Configuring persistent storage:** While containers are ephemeral by nature, many applications require persistent storage to store data that must survive container restarts or failures. We will discuss Kubernetes storage primitives, such as Persistent Volumes (PVs) and Persistent Volume Claims (PVCs), and demonstrate how to configure and use them to provide stateful applications with the necessary storage resources.

- **Implementing rolling updates and rollbacks:** Ensuring zero-downtime updates and the ability to roll back to a previous version of an application are crucial for maintaining high availability and minimizing the impact of changes on end-users. We will explore the concept of rolling updates, how to configure update strategies in Kubernetes deployments, and how to perform rollbacks in case of issues.

- **Setting up autoscaling:** To ensure optimal application performance and resource utilization, it is essential to scale the number of running instances of an application based on demand. We will discuss Kubernetes autoscaling mechanisms, such as Horizontal Pod Autoscaling (HPA) and Cluster Autoscaling, and demonstrate how to configure and use them to dynamically adjust the number of running instances based on resource usage or custom metrics.

For instance, consider a web application that experiences varying levels of traffic throughout the day. By implementing autoscaling, the application can automatically scale the number of running instances during periods of high demand, ensuring optimal performance and efficient resource utilization. Additionally, by leveraging rolling updates, the application can be updated seamlessly without causing downtime, and any issues encountered during the update can be quickly rolled back to minimize the impact on end-users.

In conclusion, this section will provide you with a deeper understanding of advanced Kubernetes concepts and best practices, equipping you with the knowledge and skills needed to optimize your containerized applications for performance, scalability, and reliability. As you progress through this module, remember to engage with the practical examples and exercises, as they will help you develop the hands-on skills necessary to effectively manage containerized applications in a Kubernetes cluster.

Kubernetes in AWS, Microsoft Azure, and Google Cloud

Kubernetes is a popular container orchestration platform used across major cloud providers, such as AWS, Microsoft Azure, and Google Cloud, to manage, deploy, and scale containerized applications. Each cloud provider offers managed Kubernetes services that simplify the process of deploying and managing

Kubernetes clusters. In this section, we will discuss how Kubernetes is used in each of these cloud providers, providing practical examples and configuration details. We will also explain the differences and similarities between Kubernetes and Docker.

AWS:

Amazon Web Services (AWS) offers the Amazon Elastic Kubernetes Service (EKS) to manage Kubernetes clusters.

- **Amazon Elastic Kubernetes Service (EKS):** EKS is a managed Kubernetes service that simplifies the deployment, scaling, and management of Kubernetes clusters on AWS. It integrates with various AWS services, such as Elastic Load Balancing, Amazon RDS, and AWS Identity and Access Management (IAM).

  Practical Example: To deploy a Kubernetes cluster with Amazon EKS, follow these steps:

  - Create an Amazon EKS cluster using the AWS Management Console or AWS CLI.
  - Configure your kubectl CLI tool to use the EKS cluster.
  - Deploy your applications using Kubernetes manifest files that define your desired application state.

Microsoft Azure:

Microsoft Azure provides Azure Kubernetes Service (AKS) to manage Kubernetes clusters.

- **Azure Kubernetes Service (AKS):** AKS is a managed Kubernetes service that simplifies the deployment, scaling, and management of Kubernetes clusters on Azure. It integrates with Azure services, such as Azure Active Directory, Azure Monitor, and Azure Virtual Networks.

Practical Example: To deploy a Kubernetes cluster with Azure Kubernetes Service, follow these steps:

- Create an Azure Kubernetes Service cluster using the Azure Portal or Azure CLI.
- Configure your kubectl CLI tool to use the AKS cluster.
- Deploy your applications using Kubernetes manifest files that define your desired application state.

Google Cloud:

Google Cloud offers Google Kubernetes Engine (GKE) to manage Kubernetes clusters.

- **Google Kubernetes Engine (GKE):** GKE is a managed Kubernetes service that simplifies the deployment, scaling, and management of Kubernetes clusters on Google Cloud. It integrates with various Google Cloud services, such as Google Cloud Storage, Google Cloud SQL, and Google Cloud Load Balancing.

  Practical Example: To deploy a Kubernetes cluster with Google Kubernetes Engine, follow these steps:

  - Create a Google Kubernetes Engine cluster using the Google Cloud Console or gcloud CLI.
  - Configure your kubectl CLI tool to use the GKE cluster.
  - Deploy your applications using Kubernetes manifest files that define your desired application state.

Differences and Similarities between Kubernetes and Docker:

- Docker is a platform that allows you to build, package, and run containerized applications. It focuses on the container level, enabling developers to create and manage containers with their application code and dependencies. Docker simplifies the process of deploying applications consistently across different environments by using container images.

- Kubernetes, on the other hand, is a container orchestration platform that focuses on the management, deployment, scaling, and availability of containerized applications. Kubernetes is not limited to Docker containers but can also manage other container runtimes, such as containerd and CRI-O. It provides advanced features like load balancing, rolling updates, rollbacks, and autoscaling, which are essential for managing large-scale containerized applications.
- While both Docker and Kubernetes deal with containerized applications, they serve different purposes. Docker focuses on creating and managing individual containers, whereas Kubernetes is designed to orchestrate and manage clusters of containers. In other words, Docker is responsible for running containers, while Kubernetes handles the orchestration and management of containerized applications at scale.

In conclusion, Kubernetes is widely supported across major cloud providers, such as AWS, Microsoft Azure, and Google Cloud. Each provider offers managed Kubernetes services that simplify the process of deploying and managing Kubernetes clusters. By understanding the different services and their usage, you can choose the most suitable approach for deploying and managing your Kubernetes clusters in the cloud.

Docker and Kubernetes serve different but complementary purposes in the world of containerization. While Docker is focused on building, packaging, and running containerized applications, Kubernetes is designed to orchestrate and manage the deployment, scaling, and availability of these applications. Together, they provide a powerful ecosystem for creating, deploying, and managing modern, scalable, and resilient containerized applications. By leveraging the managed Kubernetes services offered by major cloud providers, you can further simplify the process of deploying and managing containerized applications in the cloud.

Serverless Computing (AWS Lambda, Azure Functions, Google Cloud Functions)

In this section, you'll be introduced to the concept of serverless computing, its benefits, and how it enables you to build and run applications without worrying about infrastructure management and scaling. Serverless computing is an innovative cloud-based execution model that allows you to focus on writing code and developing the core functionality of your applications while leaving the provisioning, maintenance, and scaling of resources to the cloud provider.

The key benefits of serverless computing include:

- **Cost efficiency:** With serverless computing, you only pay for the compute resources consumed during the execution of your code. This pay-as-you-go model can lead to significant cost savings compared to traditional cloud computing models that require pre-allocated resources.
- **Scalability:** Serverless platforms automatically scale your applications based on demand. This means that your applications can seamlessly handle varying levels of traffic without manual intervention, ensuring consistent performance and responsiveness.
- **Reduced operational overhead:** Since the cloud provider takes care of infrastructure management, you don't need to spend time and effort on tasks such as server provisioning, patching, and maintenance. This allows you to focus on your application's core functionality and features.
- **Faster time-to-market:** Serverless computing simplifies the development and deployment process, enabling you to quickly build, test, and deploy new features and

applications without worrying about the underlying infrastructure.

Serverless applications are typically built around an event-driven architecture, where your code is executed in response to specific events or triggers, such as changes in data, user actions, or messages from other services. This execution model enables you to create highly responsive, scalable, and efficient applications that can handle a wide range of use cases, such as data processing, real-time analytics, chatbots, and IoT applications.

Understanding the concept of serverless computing, its benefits, and execution model, will allow you to be well-equipped to leverage this innovative approach to build and deploy powerful, scalable, and cost-effective applications in the cloud.

AWS Lambda

In this section, you will learn the fundamentals of AWS Lambda, a serverless compute service that allows you to run your code without provisioning or managing servers. With AWS Lambda, you can build and deploy scalable, event-driven applications while AWS takes care of the underlying infrastructure.

Creating a Lambda function involves writing your code (in a language supported by the AWS Lambda runtime environment), packaging it with any required dependencies, and uploading it to the AWS Lambda service. You can create Lambda functions using the AWS Management Console, AWS CLI, or SDKs.

To deploy a Lambda function, follow these steps:

- **Write your code:** Choose a supported runtime environment (such as Node.js, Python, or Java) and write the code for your Lambda function.
- **Package your function:** Bundle your code and any required dependencies into a deployment package (ZIP or JAR file).

- **Upload your function:** Upload the deployment package to AWS Lambda using the AWS Management Console, AWS CLI, or SDKs.

Lambda functions can be triggered using various event sources, such as:

- **Amazon S3:** When an object is created or deleted in an S3 bucket, a Lambda function can be triggered to process the object.
- **Amazon API Gateway:** Create an API that maps to a Lambda function, allowing you to build serverless RESTful APIs.
- **Amazon SNS or SQS:** Lambda functions can be triggered by messages published to SNS topics or added to SQS queues.
- **CloudWatch Events:** Schedule Lambda functions to run on a specific time interval or in response to specific AWS events.

Lambda functions can also be integrated with other AWS services, such as:

- **Amazon DynamoDB:** Read or write data to DynamoDB tables in response to Lambda function invocations.
- **AWS Step Functions:** Orchestrate multiple Lambda functions in a serverless workflow.
- **Amazon Kinesis:** Process streaming data in real-time using Lambda functions.

For example, you can create a Lambda function that automatically generates thumbnails for images uploaded to an S3 bucket. When an image is uploaded, an S3 event triggers the Lambda function, which reads the image, creates a thumbnail, and stores the thumbnail back to another S3 bucket.

Understanding the fundamentals of AWS Lambda will allow you to be able to create, deploy, and manage serverless applications on AWS, leveraging various event sources and integrating with other AWS services to build powerful and scalable solutions.

Azure Functions

In this section, you will explore Azure Functions, a serverless compute service that enables you to run event-driven code without managing infrastructure. Azure Functions allows you to build and deploy scalable, event-driven applications on the Azure platform while abstracting the underlying infrastructure management.

Creating an Azure Function involves writing your code (in a language supported by the Azure Functions runtime, such as C#, Java, JavaScript, or Python), packaging it with any required dependencies, and uploading it to the Azure Functions service. You can create Azure Functions using the Azure Portal, Azure CLI, or SDKs.

To deploy an Azure Function, follow these steps:

- **Write your code:** Choose a supported runtime environment and write the code for your Azure Function.
- **Package your function:** Bundle your code and any required dependencies into a deployment package (ZIP file).
- **Upload your function:** Upload the deployment package to Azure Functions using the Azure Portal, Azure CLI, or SDKs.

Azure Functions can be triggered using various event sources, such as:

- **HTTP triggers:** Expose your Azure Functions as HTTP endpoints to create serverless RESTful APIs.

- **Timer triggers:** Schedule your Azure Functions to run on a specific time interval, like a cron job.
- **Blob Storage triggers:** Run your Azure Functions in response to changes in Azure Blob Storage, such as when a new file is added or an existing file is updated.
- Event Hubs or Service Bus: Trigger Azure Functions when messages are sent to Event Hubs or Service Bus queues and topics.

Azure Functions can also be integrated with other Azure services, such as:

- **Cosmos DB:** Read or write data to Cosmos DB collections in response to Azure Function invocations.
- **Azure Storage:** Store or retrieve files, such as images or documents, using Azure Blob Storage, File Storage, or Queue Storage.
- **Azure Logic Apps:** Orchestrate multiple Azure Functions in a serverless workflow.

For example, you can create an Azure Function that processes messages from an IoT device and stores the processed data in Cosmos DB. When a new message arrives in an Event Hub, the Azure Function is triggered, processes the message, and writes the resulting data to a Cosmos DB collection.

Exploring Azure Functions will allow you create, deploy, and manage serverless applications on the Azure platform, leveraging various event sources and integrating with other Azure services to build powerful and scalable solutions.

Google Cloud Functions

In this section, you will delve into Google Cloud Functions, a serverless compute platform for creating event-driven applications on Google Cloud. Google Cloud Functions allows you to build and deploy scalable applications without worrying about infrastructure management, as Google Cloud handles it for you.

Creating a Cloud Function involves writing your code (in a language supported by the Google Cloud Functions runtime, such as Node.js, Python, or Go), packaging it with any required dependencies, and uploading it to the Google Cloud Functions service. You can create Cloud Functions using the Google Cloud Console, Google Cloud CLI (gcloud), or SDKs.

To deploy a Cloud Function, follow these steps:

- **Write your code:** Choose a supported runtime environment and write the code for your Cloud Function.
- **Package your function:** Bundle your code and any required dependencies into a deployment package (ZIP file).
- **Upload your function:** Upload the deployment package to Google Cloud Functions using the Google Cloud Console, gcloud, or SDKs.

Google Cloud Functions can be triggered using various event sources, such as:

- HTTP triggers: Expose your Cloud Functions as HTTP endpoints to create serverless RESTful APIs.
- **Pub/Sub triggers:** Run your Cloud Functions in response to messages published to Google Cloud Pub/Sub topics.
- **Storage triggers:** Execute your Cloud Functions when changes occur in Google Cloud Storage, such as new file uploads or file deletions.
- **Firestore triggers:** Trigger Cloud Functions based on events in Firestore, like document creations, updates, or deletions.

Google Cloud Functions can also be integrated with other Google Cloud services, such as:

- **BigQuery:** Insert, update, or query data in BigQuery tables in response to Cloud Function invocations.
- **Datastore:** Read or write data to Datastore entities during the execution of your Cloud Function.
- **Cloud Logging:** Monitor and analyze the logs generated by your Cloud Functions using Cloud Logging.

For example, you can create a Cloud Function that generates thumbnails for images uploaded to Google Cloud Storage. When a new image is uploaded, a storage event triggers the Cloud Function, which reads the image, creates a thumbnail, and stores the thumbnail back to a different Google Cloud Storage bucket.

Delving into Google Cloud Functions will allow you create, deploy, and manage serverless applications on Google Cloud, leveraging various event sources and integrating with other Google Cloud services to build powerful and scalable solutions.

Comparing Serverless Compute Platforms

In this section, you will compare the features, pricing, and capabilities of AWS Lambda, Azure Functions, and Google Cloud Functions. By understanding the differences and similarities between these serverless platforms, you can make an informed decision on the right one for your use case.

Features:

- **AWS Lambda:** AWS Lambda supports a variety of languages, including Node.js, Python, Java, Go, .NET, Ruby, and PowerShell. It integrates seamlessly with other AWS services and can be triggered by various event sources like API Gateway, S3, DynamoDB, etc.

- **Azure Functions:** Azure Functions supports a range of languages, such as C#, F#, Node.js, Python, Java, and PowerShell. It integrates with other Azure services and can be triggered by various event sources, like HTTP requests, Azure Storage, Cosmos DB, Event Hubs, etc.

- **Google Cloud Functions:** Google Cloud Functions supports Node.js, Python, and Go. It integrates with other Google Cloud services and can be triggered by different event sources, like HTTP requests, Cloud Pub/Sub, Cloud Storage, and Firestore.

Pricing:

- **AWS Lambda:** You are charged based on the number of requests and the duration of function execution. AWS provides a free tier with 1 million requests and 400,000 GB-seconds per month.

- **Azure Functions:** Pricing is based on the number of executions, memory usage, and execution duration. Azure provides a free tier with 1 million requests and 400,000 GB-seconds per month.

- **Google Cloud Functions:** You are billed for the number of invocations, memory usage, and execution time. Google Cloud Functions offers a free tier with 2 million invocations, 400,000 GB-seconds, and 200,000 GHz-seconds per month.

Capabilities:

- **AWS Lambda:** AWS Lambda provides features like layers for sharing code and libraries, provisioned concurrency for predictable performance, and custom runtimes for using other languages or environments.

- **Azure Functions:** Azure Functions offers features like Durable Functions for orchestrating long-running workflows, deployment slots for staging and swapping, and integration with Azure DevOps and GitHub Actions.

- **Google Cloud Functions:** Google Cloud Functions provides features like VPC connectivity, Cloud Build integration for CI/CD, and support for serverless containers with Cloud Run.

When choosing the right serverless platform for your use case, consider factors like:

- **Language support:** Ensure the platform supports the programming languages you prefer to use.
- **Integration with other services:** Check if the platform integrates well with the services you are already using or plan to use.
- **Pricing model:** Evaluate the cost based on your expected usage patterns and budget constraints.
- **Features and capabilities:** Determine if the platform provides the features you need to implement your specific use case effectively.

For example, if you are heavily invested in the AWS ecosystem, AWS Lambda may be the best choice for seamless integration with other AWS services. If you are developing an application in .NET, Azure Functions could be the right fit due to its native support for .NET languages. Similarly, if you are using Google Cloud services, Google Cloud Functions would offer smooth integration and ease of use within the Google Cloud ecosystem.

Comparing serverless compute platforms will allow you to be better equipped to choose the right one that fits your needs and optimizes your serverless application development experience.

Practical: Deploying and Scaling Applications in the Cloud

In this section we will dive into the hands-on aspects of deploying, managing, and scaling containerized applications on major cloud platforms, including AWS, Microsoft Azure, and Google Cloud. Throughout this module, you will gain valuable practical experience in leveraging managed container and Kubernetes services provided by these cloud providers to streamline the deployment process and ensure optimal performance and resource utilization.

You will learn how to deploy containerized applications using managed container services, set up and manage Kubernetes clusters, and implement advanced features like load balancing and autoscaling to accommodate dynamic workloads. Additionally, you will explore monitoring and troubleshooting tools available on cloud platforms to help you identify and resolve performance issues and optimize your application deployments.

By the end of this section, you will be well-equipped with practical knowledge and hands-on experience in deploying, managing, and scaling containerized applications in the cloud, enabling you to build highly available, scalable, and resilient applications for real-world scenarios.

Deploying a containerized application using managed container services

In this part of the module, you will learn how to deploy your containerized applications on cloud platforms using managed container services such as AWS Elastic Container Service (ECS) or Fargate, Azure Container Instances (ACI), and Google Cloud Run. These services provide a simplified deployment process and abstract the underlying infrastructure management, allowing you to focus on your application code and functionality.

AWS:

- **AWS Elastic Container Service (ECS)** is a managed container service that enables you to run, stop, and manage Docker containers on a cluster. **Fargate**, a serverless compute engine for containers, works with ECS to further simplify the deployment process by removing the need to manage the underlying infrastructure. You can define your container specifications, resource requirements, and networking configurations in a task definition and launch your containers on a Fargate launch type.

  Practical Example: To deploy your containerized application using AWS ECS and Fargate, follow these steps:

  1. Create a Docker image for your application and push it to a container registry like Amazon Elastic Container Registry (ECR).

  2. Create an ECS task definition, specifying the container image, resources, and networking configurations.

  3. Launch the task using the Fargate launch type and configure the desired number of container instances.

Microsoft Azure:

- **Azure Container Instances (ACI)** is a managed service that allows you to run containerized applications on Azure without the need to manage the underlying infrastructure. You can easily deploy your containers by specifying the container image, resource requirements, and networking configurations.

Practical Example: To deploy your containerized application using Azure Container Instances, follow these steps:

1. Create a Docker image for your application and push it to a container registry like Azure Container Registry (ACR).
2. Use the Azure CLI or Azure Portal to create a container instance, specifying the container image, resources, and networking configurations.
3. Monitor and manage your container instances using the Azure Portal or Azure CLI.

Google Cloud:

- **Cloud Run** is a fully managed serverless platform that enables you to run containerized applications without worrying about the underlying infrastructure. It automatically scales your containers based on incoming traffic and supports both HTTP and gRPC services.

  Practical Example: To deploy your containerized application using Google Cloud Run, follow these steps:

  - Build and push your Docker image to Google Container Registry (GCR) or another container registry.
  - Use the Google Cloud Console or the gcloud CLI to deploy the container to Cloud Run, specifying the image and other configurations.
  - Monitor and manage your container instances using the Google Cloud Console or the gcloud CLI.

Leveraging these managed container services will allow you quickly and easily deploy your containerized applications in the cloud while focusing on your application's core functionality rather than managing the underlying infrastructure.

Setting up a managed Kubernetes cluster

In this section, you will explore how to create and configure a managed Kubernetes cluster on AWS Elastic Kubernetes Service (EKS), Azure Kubernetes Service (AKS), or Google Kubernetes Engine (GKE). Managed Kubernetes services offer numerous benefits in terms of cluster management, security, and scalability, allowing you to focus on developing and deploying your applications without the need to manage the underlying Kubernetes infrastructure.

AWS:

- **AWS Elastic Kubernetes Service (EKS)** is a managed Kubernetes service that simplifies the process of running Kubernetes on AWS. It automates the deployment, scaling, and management of Kubernetes clusters while maintaining compatibility with existing Kubernetes tools and plugins.

  Practical Example: To create and configure a managed Kubernetes cluster on AWS EKS, follow these steps:

  1. Use the AWS Management Console or the AWS CLI to create an EKS cluster, specifying the desired Kubernetes version, VPC, and subnet configurations.
  2. Configure the Kubernetes CLI (kubectl) to interact with your EKS cluster.
  3. Deploy your applications using Kubernetes manifests, and manage your cluster using kubectl and the Kubernetes Dashboard.

Microsoft Azure:

- **Azure Kubernetes Service (AKS)** is a managed Kubernetes service that simplifies the deployment, scaling, and management of Kubernetes clusters on Azure. It offers seamless integration with other Azure services and tools and provides built-in support for monitoring and security features.

  Practical Example: To create and configure a managed Kubernetes cluster on Azure AKS, follow these steps:

  1. Use the Azure Portal or the Azure CLI to create an AKS cluster, specifying the desired Kubernetes version, resource group, and networking configurations.

  2. Configure kubectl to interact with your AKS cluster.

  3. Deploy your applications using Kubernetes manifests, and manage your cluster using kubectl and the Kubernetes Dashboard.

Google Cloud:

- **Google Kubernetes Engine (GKE)** is a managed Kubernetes service that simplifies the deployment, scaling, and management of Kubernetes clusters on Google Cloud. GKE provides a highly reliable, secure, and scalable environment for running your containerized applications.

  Practical Example: To create and configure a managed Kubernetes cluster on Google GKE, follow these steps:

  1. Use the Google Cloud Console or the gcloud CLI to create a GKE cluster, specifying the desired Kubernetes version, zone or region, and networking configurations.

  2. Configure kubectl to interact with your GKE cluster.

3. Deploy your applications using Kubernetes manifests, and manage your cluster using kubectl and the Kubernetes Dashboard.

Using managed Kubernetes services like EKS, AKS, or GKE, will allow you take advantage of their simplified cluster management, security features, and scalability capabilities, allowing you to focus on deploying and managing your containerized applications in the cloud.

Implementing load balancing and autoscaling

In this section, you will learn how to set up load balancing and autoscaling for your cloud-deployed applications using native cloud services and Kubernetes features. You'll explore how to use AWS Elastic Load Balancing, Azure Load Balancer, and Google Cloud Load Balancing, as well as Kubernetes Ingress controllers and the Horizontal Pod Autoscaler (HPA) to distribute traffic and automatically scale your applications based on demand.

AWS:

- **AWS Elastic Load Balancing (ELB)** offers three types of load balancers: Application Load Balancer (ALB), Network Load Balancer (NLB), and Classic Load Balancer (CLB). These load balancers distribute incoming traffic across multiple targets, such as EC2 instances or containers running in an ECS or EKS cluster.

  Practical Example: To set up load balancing for your application on AWS, follow these steps:

  1. Create an ALB, NLB, or CLB using the AWS Management Console or the AWS CLI.
  2. Configure the load balancer to route traffic to your application's backend instances or containers.

3. Update your application's DNS records to point to the load balancer's hostname or IP address.

Microsoft Azure:

- **Azure Load Balancer** provides both Layer-4 (TCP/UDP) and Layer-7 (HTTP/HTTPS) load balancing capabilities. It distributes incoming traffic across multiple VMs or containers running in an AKS cluster.

  Practical Example: To set up load balancing for your application on Azure, follow these steps:

  - Create an Azure Load Balancer using the Azure Portal or the Azure CLI.
  - Configure the load balancer to route traffic to your application's backend VMs or containers.
  - Update your application's DNS records to point to the load balancer's IP address or hostname.

Google Cloud:

- **Google Cloud Load Balancing** offers various types of load balancers, including HTTP(S), TCP/SSL Proxy, and Network TCP/UDP load balancers. These load balancers distribute incoming traffic across multiple instances or containers running in a GKE cluster.

  Practical Example: To set up load balancing for your application on Google Cloud, follow these steps:

  1. Create a Google Cloud Load Balancer using the Google Cloud Console or the gcloud CLI.
  2. Configure the load balancer to route traffic to your application's backend instances or containers.
  3. Update your application's DNS records to point to the load balancer's IP address or hostname.

Kubernetes Ingress and Ingress Controllers:

- Kubernetes Ingress is an API object that defines how external traffic should be routed to services within a cluster. Ingress controllers, such as Nginx, HAProxy, or Traefik, implement the Ingress rules and provide load balancing capabilities.

  Practical Example: To set up load balancing for your application using Kubernetes Ingress, follow these steps:

  1. Deploy an Ingress controller in your Kubernetes cluster.
  2. Create an Ingress resource that defines the routing rules for your application.
  3. Update your application's DNS records to point to the Ingress controller's IP address or hostname.

Kubernetes Horizontal Pod Autoscaler (HPA):

- HPA automatically scales the number of pods in a Kubernetes deployment or replication controller based on observed CPU utilization or custom metrics.

  Practical Example: To set up autoscaling for your application using Kubernetes HPA, follow these steps:

  1. Deploy the Metrics Server in your Kubernetes cluster to collect resource usage metrics.
  2. Create an HPA resource that specifies the target CPU utilization percentage and the minimum and maximum number of pods for your deployment.
  3. Monitor the HPA's scaling actions and performance using kubectl or the Kubernetes Dashboard.

Implementing load balancing and autoscaling will allow you to ensure that your cloud-deployed applications are highly available and can efficiently handle varying levels of traffic. Leveraging native cloud services like AWS Elastic Load Balancing, Azure Load Balancer, and Google Cloud Load Balancing, as well as Kubernetes features like Ingress controllers and Horizontal Pod Autoscaler, you can distribute incoming traffic evenly across your application instances and automatically scale the number of running instances or pods based on demand. This approach not only improves your application's performance and reliability but also helps optimize resource utilization and cost management in the cloud.

Monitoring and troubleshooting deployed applications

In this section, you will gain practical knowledge of monitoring and troubleshooting tools available in cloud platforms, such as AWS CloudWatch, Azure Monitor, and Google Cloud Operations Suite. These tools play a crucial role in identifying performance issues, bottlenecks, and potential areas for optimization, ensuring your deployed applications are running smoothly and efficiently.

AWS:

- **AWS CloudWatch** is a monitoring and observability service that collects and analyzes metrics, logs, and events from your AWS resources, applications, and services. It enables you to set alarms and create dashboards to visualize and analyze the performance of your applications.

  Practical Example: To monitor and troubleshoot your application on AWS, follow these steps:

  1. Configure your resources, applications, and services to send metrics and logs to CloudWatch.
  2. Use CloudWatch Metrics to visualize and analyze performance data.

3. Set up CloudWatch Alarms to receive notifications when specific thresholds are breached.

4. Use CloudWatch Logs Insights to search and analyze log data for troubleshooting and optimization.

Microsoft Azure:

- **Azure Monitor** is a comprehensive monitoring solution that collects and analyzes telemetry data from your Azure resources, applications, and services. It helps you understand their performance, identify trends, and detect issues.

  Practical Example: To monitor and troubleshoot your application on Azure, follow these steps:

  1. Configure your resources, applications, and services to send metrics and logs to Azure Monitor.

  2. Use Azure Monitor Metrics Explorer to visualize and analyze performance data.

  3. Set up Azure Monitor Alerts to receive notifications when specific thresholds are breached.

  4. Use Azure Monitor Log Analytics to search and analyze log data for troubleshooting and optimization.

Google Cloud:

- **Google Cloud Operations Suite**, formerly known as Stackdriver, is a suite of monitoring, logging, and diagnostics tools that help you gain insights into the performance, availability, and health of your applications running on Google Cloud.

Practical Example: To monitor and troubleshoot your application on Google Cloud, follow these steps:

1. Configure your resources, applications, and services to send metrics and logs to Google Cloud Operations Suite.
2. Use Google Cloud Monitoring to visualize and analyze performance data.
3. Set up Google Cloud Alerting to receive notifications when specific thresholds are breached.
4. Use Google Cloud Logging to search and analyze log data for troubleshooting and optimization.

Utilizing monitoring and troubleshooting tools like AWS CloudWatch, Azure Monitor, and Google Cloud Operations Suite, will allow you proactively identify and resolve performance issues, bottlenecks, and potential areas for optimization in your cloud-deployed applications. These tools not only help ensure your applications are running efficiently but also allow you to make data-driven decisions for continuous improvement and cost management in the cloud.

Updating and rolling back applications

In this section, you will learn how to perform zero-downtime updates and rollbacks for your cloud-deployed applications using Kubernetes rolling updates and deployments, as well as cloud-native services like AWS CodeDeploy or Azure DevOps Pipelines. These techniques ensure that your applications remain available and responsive during updates, minimizing the risk of disruptions and downtime.

Kubernetes Rolling Updates and Deployments:

- Kubernetes deployments support rolling updates, which enable you to update your application's container images with minimal impact on availability. During a rolling update, Kubernetes gradually replaces old replicas with new ones, ensuring that at least a certain number of replicas are always available.

  Practical Example: To perform a rolling update of your application in Kubernetes, follow these steps:

  1. Update the container image in your deployment manifest or use the kubectl set image command.
  2. Apply the updated manifest or run the kubectl rollout command to initiate the rolling update.
  3. Monitor the progress of the update using kubectl rollout status
  4. If needed, use kubectl rollout undo to roll back the update and revert to the previous version.

AWS:

- **AWS CodeDeploy** is a deployment service that automates application updates across your infrastructure. It supports various deployment strategies, including rolling updates and blue/green deployments, which help minimize downtime and reduce the risk of deployment failures.

  Practical Example: To update your application using AWS CodeDeploy, follow these steps:

  1. Create a new application revision and upload it to an S3 bucket or a GitHub repository.
  2. Create or update a CodeDeploy deployment group that specifies the target instances or containers and the desired deployment strategy.

3. Start a new deployment using the AWS Management Console, AWS CLI, or SDKs.

4. Monitor the deployment progress, and if needed, use the CodeDeploy console or CLI to roll back the update.

Microsoft Azure:

- **Azure DevOps Pipelines** is a continuous integration and continuous delivery (CI/CD) service that automates the build, test, and deployment of your applications. It supports various deployment strategies, including rolling updates and blue/green deployments, to ensure minimal downtime during updates.

  Practical Example: To update your application using Azure DevOps Pipelines, follow these steps:

  1. Create a new build pipeline that compiles your application and creates a deployable package.

  2. Create a release pipeline that deploys the package to your target infrastructure using the desired deployment strategy.

  3. Trigger the build and release pipelines manually or automatically when changes are pushed to your source code repository.

  4. Monitor the deployment progress in the Azure DevOps console, and if needed, use the console or REST API to roll back the update.

Mastering techniques such as Kubernetes rolling updates and deployments, AWS CodeDeploy, and Azure DevOps Pipelines, will allow you confidently update and roll back your cloud-deployed applications with minimal impact on availability. These approaches not only help ensure a smooth update process but also enable you to quickly revert to a previous version in case of issues or failures, thus maintaining the reliability and performance of your applications.

Module 5 Exam Questions

1. What is the main benefit of containerization?

a) Increased server load

b) Inconsistent application execution

c) Faster development and deployment

d) Reduced security

2. Which of the following is a containerization platform?

a) Docker

b) Kubernetes

c) AWS Lambda

d) Azure Functions

3. What is the primary role of Kubernetes in container orchestration?

a) To build Docker images

b) To automate deployment, scaling, and management of containerized applications

c) To execute serverless functions

d) To monitor application performance

4. Which of the following cloud providers offers a managed Kubernetes service?

a) AWS

b) Microsoft Azure

c) Google Cloud

d) All of the above

5. What is a key advantage of serverless computing?

a) Eliminates the need for infrastructure management

b) Requires manual scaling

c) Limited language support

d) Reduced security

6. Which of the following is a serverless compute service?

a) Docker

b) Kubernetes

c) AWS Lambda

d) Google Kubernetes Engine

7. What is the primary pricing factor for serverless platforms like AWS Lambda, Azure Functions, and Google Cloud Functions?

a) Number of virtual machines

b) Amount of storage used

c) Number of requests and execution duration

d) Fixed monthly fee

8. Which cloud provider offers the Elastic Kubernetes Service (EKS)?

a) AWS

b) Microsoft Azure

c) Google Cloud

d) None of the above

9. Which of the following is an event source for AWS Lambda?

a) API Gateway

b) Azure Storage

c) Google Cloud Pub/Sub

d) None of the above

10. Which serverless compute service supports the F# programming language?

a) AWS Lambda

b) Azure Functions

c) Google Cloud Functions

d) None of the above

11. What is a key component of a Kubernetes cluster?

a) Dockerfile

b) Pod

c) Container

d) Lambda function

12. What is the purpose of a Kubernetes Ingress resource?

a) To manage application accessibility and load balancing

b) To store data persistently

c) To deploy serverless functions

d) To monitor application performance

13. Which of the following can trigger an Azure Function?

a) HTTP request

b) AWS S3 bucket

c) Google Cloud Storage

d) None of the above

14. What do Docker images consist of?

a) Only the source code

b) The application, its dependencies, and the runtime environment

c) Kubernetes configuration files

d) Serverless function code

15. What is the purpose of AWS CodeDeploy?

a) Container orchestration

b) Zero-downtime updates and rollbacks for applications

c) Serverless function deployment

d) Storing data in a database

16. What is the maximum number of invocations per month offered in the Google Cloud Functions free tier?

a) 1 million

b) 2 million

c) 5 million

d) 10 million

17. What does the Horizontal Pod Autoscaler (HPA) do in Kubernetes?

a) Automatically scales the number of Pods based on resource utilization

b) Stores data persistently across Pods

c) Manages the execution of serverless functions

d) Provides load balancing for containerized applications

18. Which tool can be used to monitor deployed applications in AWS?

a) AWS CloudWatch

b) Azure Monitor

c) Google Cloud Operations Suite

d) None of the above

19. Which cloud provider offers Azure Container Instances (ACI)?

a) AWS

b) Microsoft Azure

c) Google Cloud

d) None of the above

20. Which of the following is a key benefit of using managed Kubernetes services like AWS EKS, Azure AKS, and Google GKE?

a) Manual cluster management

b) Decreased security

c) Limited scalability

d) Simplified cluster management and security

Answers:

1. c
2. b
3. d
4. a
5. a
6. b
7. c
8. a
9. d
10. b
11. a
12. c
13. b
14. d
15. c
16. a
17. d
18. b
19. a
20. d

Module 6. Cloud Databases and Data Management

Introduction to Databases and Data Management

In today's digital world, data plays a critical role in the success of any organization. Businesses are increasingly looking to harness the power of their data to drive innovation, enhance customer experiences, and make data-driven decisions. As organizations continue to generate and process vast amounts of data, managing it effectively and securely becomes increasingly important. This is where cloud databases and data management come into play.

Cloud databases are a modern approach to managing data that leverages the power of cloud computing to deliver scalable, cost-effective, and reliable database solutions. They provide an alternative to traditional on-premises databases, allowing businesses to store, process, and manage their data in a remote, secure environment. Cloud databases offer numerous benefits, such as reduced IT infrastructure costs, simplified management, and the ability to scale resources up or down as needed to meet fluctuating demands.

Data management in the cloud involves a wide range of practices, tools, and techniques aimed at ensuring that data is stored, processed, and utilized effectively, securely, and in compliance with relevant regulations. This includes tasks such as data backup, disaster recovery, monitoring, and optimization, all of which contribute to the overall performance and resilience of the cloud database system.

A practical example of a cloud database in action can be seen in an e-commerce company. With thousands of transactions happening every day, the company needs a robust and scalable database to store customer data, order information, and inventory details. By leveraging a cloud-based database solution, the e-commerce company can easily scale its resources to accommodate traffic spikes during peak shopping seasons and ensure a seamless experience for its customers. Additionally, the company can focus on its core business operations, knowing that its data is securely managed and backed up in the cloud.

This module, "Cloud Databases and Data Management," will provide a comprehensive overview of the various types of cloud databases, the selection criteria for choosing the right database solution for your organization, and the essential strategies for data backup and disaster recovery. You will also gain hands-on experience in setting up and managing cloud databases, enabling you to put the concepts you learn into practice. By the end of this module, you will have a solid foundation in cloud databases and data management, empowering you to make informed decisions and implement best practices for your organization's data needs.

Database Types and Selection Criteria

Selecting the right database for your organization's needs is a critical decision that can significantly impact the performance, scalability, and maintainability of your applications. With a wide array of database types available, each with its unique features and capabilities, it's essential to understand the differences and make an informed choice based on your specific use case and requirements. This section, "Database Types and Selection Criteria," will provide an in-depth exploration of various database types and the factors to consider when selecting the right database for your needs.

In today's data-driven world, different applications have varying data management requirements. Some applications need to process structured data with complex relationships, while others deal with large volumes of unstructured or semi-structured data. Furthermore, certain applications may require real-time analytics or low-latency access to data, while others prioritize high availability and scalability.

To cater to these diverse needs, a range of database types have emerged, including relational databases, NoSQL databases, time-series databases, and in-memory databases. Each type comes with its strengths and limitations, making it crucial to identify the best fit for your application.

Throughout this section, we will discuss practical examples that illustrate the benefits and drawbacks of various database types in different scenarios. For instance, consider a financial services company that requires a database to store and analyze structured data, such as customer information and transaction records. In this case, a relational database with strong consistency and transaction support may be the best choice. On the other hand, a social media platform dealing with large volumes of user-generated content and complex relationships might benefit from a NoSQL database, which offers flexibility and horizontal scaling capabilities.

By the end of this section, you will have a comprehensive understanding of the various database types and the selection criteria to consider when choosing the right database for your organization. This knowledge will empower you to make informed decisions and ensure that your applications have a robust and efficient data management foundation.

Relational Databases (SQL)

Relational databases are a widely-used and mature database technology that relies on a schema to define the structure of your data. In a relational database, data is organized into tables, where each table represents an entity, and the relationships between entities are established through primary and foreign keys. These databases are based on SQL (Structured Query Language), a powerful and standardized language for querying and manipulating data.

If your application requires complex queries, strict data consistency, and transaction support, a relational database might be the ideal choice for you. These databases excel in scenarios where data integrity and the ability to perform complex operations on structured data are paramount. For example, if you're building a banking system, a human resources application, or a customer relationship management (CRM) platform, a relational database can provide the strong data consistency and querying capabilities needed to ensure data accuracy and meaningful insights.

Some popular relational databases include MySQL, PostgreSQL, Microsoft SQL Server, and Oracle Database. Each of these databases has its unique features and capabilities, but they all follow the relational model and support SQL as the primary query language. When selecting a relational database, consider factors such as licensing costs, community support, available tools, and the specific features you require, such as support for stored procedures, triggers, and advanced indexing options.

In a practical example, imagine you're building an e-commerce platform where you need to manage customer data, product inventory, and order details. A relational database can help you effectively store and manage this structured data, ensuring data consistency and enabling you to perform complex queries, such as generating sales reports, analyzing customer behavior, or identifying popular products. By choosing a relational database for your e-commerce platform, you'll have a solid foundation to support your data needs and ensure a reliable and efficient system.

AWS:

When using cloud platforms like AWS, Microsoft Azure, or Google Cloud, you can take advantage of their managed relational database services, which simplify database management and provide additional benefits like scalability, high availability, and automated backups.

In AWS, you can use Amazon RDS (Relational Database Service), which supports various relational databases such as MySQL, PostgreSQL, Microsoft SQL Server, Oracle Database, and Amazon Aurora. Amazon RDS takes care of time-consuming administration tasks like backups, patching, and scaling, allowing you to focus on your application. For example, if you're building a web application that manages customer orders, you can use Amazon RDS with PostgreSQL to ensure data consistency and easily perform complex queries on your order data.

Microsoft Azure:

Microsoft Azure offers Azure SQL Database, a fully managed relational database service based on the SQL Server engine. With built-in intelligence, it optimizes performance and security for your application. Azure SQL Database provides features like automatic backups, scaling, and high availability. Suppose you're creating an inventory management system. In that case, you can use Azure SQL Database to store your product information, maintain data integrity, and generate real-time reports on stock levels and sales trends.

Google Cloud:

Google Cloud provides Cloud SQL, a managed relational database service that supports MySQL, PostgreSQL, and Microsoft SQL Server. Cloud SQL simplifies database administration by handling tasks like backups, replication, and patching. It also integrates seamlessly with other Google Cloud services, such as App Engine and Cloud Functions. If you're developing an e-commerce platform, you can use Cloud SQL with MySQL to manage customer data, product catalogs, and order history, ensuring a reliable and efficient data management solution for your business.

Leveraging managed relational database services in AWS, Microsoft Azure, or Google Cloud, will allow you harness the power of SQL databases while enjoying the convenience, scalability, and reliability provided by these cloud platforms. This allows you to focus on your application's core functionality and deliver a seamless experience for your users.

NoSQL Databases

NoSQL databases have emerged as a popular alternative to relational databases, offering more flexibility in handling unstructured or semi-structured data and providing unique scaling capabilities. If your application deals with large volumes of data, complex relationships, or requires the ability to scale horizontally, a NoSQL database could be a suitable choice for you.

NoSQL databases come in various types, each designed to address specific data management challenges:

- **Document-based databases**, such as MongoDB, store data in documents (typically JSON or BSON format), allowing for flexible data modeling and easy-to-understand data structures. These databases are well-suited for applications that deal with diverse data, such as content management systems, event logging, or user profiles.

- **Key-value databases**, like Redis, focus on simple key-value pairs, enabling fast and efficient access to data. These databases are ideal for caching, session management, or storing configuration data, where the primary requirement is high-speed data retrieval.
- **Column-family databases**, including Apache Cassandra, store data in columns rather than rows, making them particularly efficient for read and write operations on large data sets. These databases are often used in applications with heavy write loads or requiring high availability, such as analytics platforms, IoT data processing, or recommendation engines.
- **Graph-based databases**, such as Neo4j, are designed to represent and query complex relationships between entities in a graph structure. They excel in scenarios where navigating and analyzing relationships are essential, like social networks, fraud detection systems, or knowledge graphs.

When choosing a NoSQL database, consider the specific data structure and scaling requirements of your application, as well as the database's maturity, community support, and available tools. For example, if you're building a real-time analytics dashboard that aggregates data from multiple sources, a document-based database like MongoDB might be ideal for handling diverse data types and providing flexibility in data modeling. On the other hand, if your application involves analyzing complex relationships between users, such as a social networking platform, a graph-based database like Neo4j would be a more appropriate choice.

Selecting the right NoSQL database for your needs will allow you effectively manage your data, optimize performance, and ensure the scalability and flexibility required for your application's success.

AWS:

AWS, Microsoft Azure, and Google Cloud each provide managed NoSQL database services that cater to various data management needs, such as document-based, key-value, column-family, and graph-based databases.

In AWS, you can use Amazon DynamoDB, a managed key-value and document-based NoSQL database service that delivers single-digit millisecond performance at any scale. DynamoDB is suitable for applications that require low-latency data access, such as gaming leaderboards or real-time analytics. If you're building a content management system with diverse data types, you can use DynamoDB's document-based capabilities to store and query your data efficiently.

Microsoft Azure:

Microsoft Azure offers Azure Cosmos DB, a globally distributed, multi-model NoSQL database service that supports document, key-value, graph, and column-family data models. With its turnkey global distribution, Cosmos DB is ideal for applications with users distributed across the world, such as a global social network or an e-commerce website. For instance, if you're developing a recommendation engine that analyzes user behavior, you can utilize Cosmos DB's graph capabilities to model and query complex relationships between users and products.

Google Cloud:

Google Cloud provides Cloud Firestore, a flexible, scalable, and managed document-based NoSQL database designed for mobile and web applications. Firestore simplifies data synchronization and querying, making it a great fit for applications that require real-time updates, like chat applications or collaborative tools. If you're building an IoT application that collects data from various sensors, you can use Firestore to store and query your semi-structured data, allowing you to analyze trends and generate insights.

Using managed NoSQL database services in AWS, Microsoft Azure, or Google Cloud, will allow you take advantage of their flexibility, scalability, and performance to address the unique data management requirements of your application. These cloud platforms provide you with a robust foundation for managing diverse data types while reducing the operational overhead typically associated with database administration.

Time-Series Databases

Time-series databases are specifically designed for handling time-series data, which consists of time-stamped data points collected at regular intervals or in a sequential manner. If your application requires the management and analysis of time-series data, such as monitoring IoT devices, tracking financial data, or measuring application performance metrics, a time-series database can be an ideal choice for you.

These databases are optimized for efficient storage and retrieval of time-series data, allowing you to perform advanced queries, aggregations, and analysis on large volumes of time-stamped data. They typically offer built-in features for handling time-based queries, data retention policies, and data compression, making them well-suited for managing historical data and real-time data streams.

Some popular time-series databases include InfluxDB, TimescaleDB, and OpenTSDB. Each of these databases has its unique features and capabilities, but they all focus on providing efficient and scalable solutions for managing time-series data. When selecting a time-series database, consider factors such as performance, scalability, data ingestion rates, query capabilities, and ease of integration with your existing infrastructure and tools.

For a practical example, imagine you're developing an IoT-based smart home system where numerous sensors collect data on temperature, humidity, and energy usage. A time-series database can help you effectively store and analyze this data, allowing you to generate insights, identify trends, and optimize energy consumption. By choosing a time-series database for your smart home system, you'll have a powerful foundation for managing and analyzing the time-stamped data generated by your IoT devices, enabling you to make data-driven decisions and improve the overall efficiency of your system.

AWS, Microsoft Azure, and Google Cloud offer managed time-series database services that simplify database management while providing the performance and scalability needed for time-series data workloads.

AWS:

In AWS, you can use Amazon Timestream, a managed time-series database designed for collecting, storing, and processing time-series data. Timestream provides fast ingestion and querying capabilities while optimizing storage and cost efficiency. If you're building a monitoring system for industrial equipment, you can use Amazon Timestream to store and analyze sensor data, enabling you to identify patterns and predict potential issues before they occur.

Microsoft Azure:

Microsoft Azure offers Azure Time Series Insights, a fully managed, multi-layered time-series database and analytics service. Time Series Insights provides powerful querying, visualization, and integration capabilities that allow you to gain insights into your time-series data. For example, if you're developing an application that tracks energy consumption across multiple buildings, you can use Azure Time Series Insights to analyze historical data and identify trends, helping you optimize energy usage and reduce costs.

Google Cloud:

Google Cloud provides Cloud IoT Core with Pub/Sub and Bigtable, a combination of services that can be used to build a custom time-series database solution. Cloud IoT Core is a fully managed service for connecting and managing IoT devices, while Pub/Sub enables real-time messaging between devices and services. Bigtable is a managed NoSQL database service designed for low-latency, high-throughput data workloads, making it suitable for storing and querying time-series data. If you're developing a smart city application that monitors traffic patterns, you can use these Google Cloud services to collect, store, and analyze time-stamped data from traffic sensors, providing insights for traffic management and urban planning.

Leveraging managed time-series database services in AWS, Microsoft Azure, or Google Cloud, will allow you efficiently store and analyze time-stamped data while benefiting from the scalability, performance, and reliability provided by these cloud platforms. This enables you to focus on deriving insights and building data-driven solutions to improve your application's performance and overall user experience.

In-Memory Databases

In-memory databases store data in memory instead of on disk, allowing for low-latency access and high performance, particularly for read-heavy or write-heavy workloads. These databases are an excellent choice for applications that require real-time analytics, caching, and session management.

One popular in-memory database is Redis, which supports data structures such as strings, hashes, lists, sets, and sorted sets. You can use Redis to implement caching for your web application, speeding up response times and reducing the load on your primary database. Redis is also useful for managing user sessions, ensuring fast and consistent access to session data across multiple instances of your application.

Memcached is another in-memory database designed for simplicity and performance. It is primarily used as a caching system for web applications, helping to reduce database load and improve response times. By storing frequently accessed data in memory, Memcached can serve requests faster than retrieving the same data from disk-based storage systems.

Amazon ElastiCache is a managed in-memory data store and cache service that supports both Redis and Memcached. With ElastiCache, you can easily deploy, run, and scale in-memory databases in the AWS cloud, benefiting from improved performance, scalability, and reliability. For instance, if you're building an e-commerce website, you can use ElastiCache to cache product details, shopping cart data, and user preferences, ensuring a fast and smooth user experience.

Incorporating in-memory databases like Redis, Memcached, or Amazon ElastiCache into your application architecture, will allow you achieve low-latency data access and high performance, enhancing the responsiveness and efficiency of your application. This allows you to deliver an improved user experience, enabling your application to better serve the needs of your users.

AWS, Microsoft Azure, and Google Cloud offer managed in-memory database services that simplify database management and ensure scalability and reliability for your applications.

AWS:

In AWS, you can use Amazon ElastiCache, a managed in-memory data store and cache service that supports both Redis and Memcached. With ElastiCache, you can easily deploy, run, and scale in-memory databases in the AWS cloud. For example, if you're building a web application that requires fast access to user data, you can use ElastiCache with Redis to cache frequently accessed information, improving response times and reducing the load on your primary database.

Microsoft Azure:

Microsoft Azure offers Azure Cache for Redis, a managed in-memory data store based on the popular Redis database. Azure Cache for Redis provides high throughput and low-latency data access, making it a suitable choice for caching, session management, and real-time analytics. If you're developing a content delivery platform, you can use Azure Cache for Redis to store and serve frequently requested content, ensuring a smooth and responsive user experience.

Google Cloud:

Google Cloud provides Cloud Memorystore, a fully managed in-memory data store service that supports both Redis and Memcached. Cloud Memorystore is designed for applications that require a fast, scalable, and reliable caching solution. For instance, if you're building a recommendation engine that needs to access user preferences quickly, you can use Cloud Memorystore with Redis to store and retrieve this data with low latency, allowing you to generate personalized recommendations in real-time.

Leveraging managed in-memory database services in AWS, Microsoft Azure, or Google Cloud, will allow you achieve low-latency data access and high performance for your applications, enhancing their responsiveness and efficiency. These cloud platforms provide you with a robust foundation for managing in-memory data, allowing you to focus on building and optimizing your application's core functionality.

Selection Criteria

When it comes to choosing a cloud database, several factors can influence your decision. To select the most suitable database type and provider for your application, carefully evaluate your specific use case and requirements.

- **Data structure and complexity:** Consider the nature of your data, whether it is structured, semi-structured, or unstructured. Depending on your data model, you may choose a relational database like MySQL for structured data, a NoSQL database like MongoDB for semi-structured data, or a time-series database like InfluxDB for time-stamped data.

- **Scalability requirements:** Assess how your application might grow over time and the scalability needs associated with that growth. For instance, if you expect a large increase in users or data volume, you may prefer a database that can scale horizontally, like Amazon DynamoDB or Google Cloud Firestore.
- **Performance and latency needs:** Determine the performance requirements of your application, including latency and throughput. For applications that require real-time analytics or low-latency data access, an in-memory database like Redis or Amazon ElastiCache may be more suitable.
- **Consistency and transaction support:** Evaluate the level of data consistency and transaction support required by your application. If you need strong consistency guarantees and complex transactional capabilities, you may prefer a relational database like PostgreSQL or Microsoft SQL Server.
- **Ease of management:** Consider the level of management and maintenance you're willing to invest in your database. Managed database services, like AWS RDS, Azure SQL Database, or Google Cloud SQL, can offload many administrative tasks, such as backups, scaling, and patching, allowing you to focus on your application's core functionality.
- **Cost:** Analyze the costs associated with different database types and providers, including data storage, data transfer, and compute resources. Be sure to consider the total cost

of ownership, including operational costs, licensing fees, and support costs.

Considering these factors will allow you make an informed decision about the right cloud database for your application. For example, if you're building a financial analytics platform that requires complex queries and strong consistency, you might choose a relational database like Microsoft SQL Server on Azure. Alternatively, if you're developing a social media application with high write volume and diverse data types, a NoSQL database like Google Cloud Firestore may be a better fit.

Data Backup and Disaster Recovery Strategies

In today's digital landscape, data has become one of the most valuable assets for any organization. Ensuring the integrity and availability of this data is crucial, which is why implementing robust data backup and disaster recovery strategies is essential. These strategies not only protect your data from accidental loss, system failures, or disasters, but also provide a reliable foundation for restoring your application to a functional state in case of unexpected events.

Data backup and disaster recovery strategies involve creating and managing copies of your data, implementing recovery plans, and validating the effectiveness of these plans. By adopting a comprehensive approach to data protection, you can minimize downtime, reduce the risk of data loss, and maintain business continuity in the face of potential threats.

For example, consider an e-commerce platform that handles a vast amount of customer information and transaction data. The loss of such data could have severe consequences, from lost revenue to reputational damage. By establishing a well-defined backup and disaster recovery strategy, the platform's operators can ensure that data is securely stored, recoverable, and compliant with applicable regulations, thus preserving the platform's functionality and customer trust.

In this section, we will explore the essential aspects of data backup and disaster recovery strategies, including backup types, storage options, recovery planning, testing, and automation. By understanding these concepts and implementing them effectively, you can safeguard your data and ensure the resilience of your application in the face of unforeseen challenges.

Backup Types and Frequency

To protect your data effectively, it's essential to understand the different types of backups and choose the appropriate backup frequency for your application, based on its criticality and data change rate.

- **Full Backups:** A full backup involves creating a complete copy of your entire dataset. While full backups provide the most comprehensive protection, they can be time-consuming and require more storage space. As an example, if you have a small application with relatively static data, performing a full backup once a week may suffice.

- **Incremental Backups:** Incremental backups store only the changes made to the data since the last backup, whether it was a full or incremental backup. This approach saves time and storage space, but recovery may take longer since multiple incremental backups may need to be applied sequentially. For instance, if your application experiences frequent data changes, you could perform a full backup weekly and incremental backups daily to balance protection and resource usage.

- **Differential Backups:** Differential backups capture the data changes since the last full backup. They require more storage space than incremental backups but enable faster recovery since only the most recent differential backup needs to be applied. In a scenario where your application has moderate data change rates, you might opt for a full backup every week and differential backups every two days.

Determining the right backup frequency for your application is crucial for maintaining data protection while optimizing resource utilization. Consider your application's criticality, the rate of data changes, and the acceptable amount of data loss in the event of a disaster when deciding on the backup frequency. By selecting the most suitable backup types and schedule, you can ensure the integrity of your data and be prepared to recover it in a timely manner when necessary.

Storage and Retention Policies

Selecting appropriate storage options and establishing data retention policies for your backup data are critical steps in ensuring the availability and compliance of your data.

- **Storage Options:** When choosing storage for your backup data, consider factors such as durability, accessibility, security, and cost. Cloud storage services like Amazon S3, Azure Blob Storage, or Google Cloud Storage offer high durability, redundancy, and scalability, making them a popular choice for storing backups. Alternatively, you can opt for offsite physical storage, like tapes or external hard drives, to protect against localized disasters. For enhanced data protection, you might consider using a combination of both cloud and offsite physical storage to create multiple copies of your backup data.

- **Retention Policies:** Determine how long your backups need to be retained to meet your application's recovery objectives and comply with any regulatory requirements. Retention policies typically specify how many backup versions to keep and the duration for which each version should be retained. For example, you might decide to retain daily backups for one week, weekly backups for one month, and monthly backups for one year. Regularly review and update your retention policies to ensure they align with your application's needs and evolving compliance requirements.

Carefully selecting suitable storage options and establishing effective retention policies will allow you to ensure that your backup data is stored securely for the necessary duration while meeting any applicable compliance requirements. This approach not only helps you maintain data availability but also supports the recovery of your application in the event of a disaster or data loss.

Disaster Recovery Planning

Developing a comprehensive disaster recovery plan is crucial for ensuring the continuity of your application and the integrity of its data in the event of a system failure, data loss, or other disasters. A well-prepared plan outlines the necessary steps to restore your application and its data, enabling you to minimize downtime and mitigate potential damages.

To create an effective disaster recovery plan, consider the following steps:

- **Identify Critical Assets:** Determine the most critical components of your application, such as databases, application servers, and user interfaces. Prioritize their recovery to ensure that essential services are restored as quickly as possible.

- **Define Recovery Objectives:** Establish the recovery point objective (RPO) and recovery time objective (RTO) for your application. RPO represents the maximum amount of data loss that can be tolerated, while RTO denotes the maximum time it should take to restore your application to a functional state.

- **Develop Recovery Procedures:** Outline the detailed steps to recover your application's critical components and data. This may include restoring backups, launching new instances of servers, or reconfiguring network settings. Document these procedures clearly, and ensure they are accessible to relevant team members.

- **Designate Responsibilities:** Assign specific roles and responsibilities to your team members for executing the recovery plan. This helps ensure that everyone is aware of their duties during a disaster and can act quickly and efficiently.

- **Communicate and Train:** Communicate the disaster recovery plan to your team and provide training on executing the plan. Regularly review and update the plan to ensure that it remains relevant and effective in the face of changing circumstances or new potential risks.

Developing a comprehensive disaster recovery plan will allow you to be better prepared to address unexpected events and ensure the continuity and resilience of your application in the face of potential threats.

Testing and Validation

To ensure the effectiveness of your backup and recovery procedures, it is essential to regularly test and validate them. By doing so, you can confirm that your data can be restored accurately and efficiently, and that your backups meet the application's recovery point objective (RPO) and recovery time objective (RTO) requirements.

- **Backup Testing:** Periodically test the integrity of your backups by performing restore operations in a controlled environment. This helps you identify any issues with the backup process, such as corrupted files or incomplete data, and take corrective action if necessary.

- **Recovery Procedure Testing:** Conduct regular recovery drills to practice and validate your documented recovery procedures. This ensures that your team is familiar with the steps to restore your application and its data, and can quickly respond in the event of a real disaster.

- **RPO and RTO Validation:** Measure the actual time it takes to restore your application and the amount of data loss during the recovery process. Compare these results with your defined RPO and RTO to ensure that your backup and recovery strategy meets the application's requirements. Adjust your strategy if needed to better align with your objectives.

- **Continuous Improvement:** Use the insights gained from testing and validation exercises to improve your backup and recovery procedures. This may include refining your recovery plan, optimizing backup frequency, or updating team training.

For example, you might perform quarterly recovery tests to simulate a data loss event and measure how long it takes to restore your application to a functional state. If the recovery time exceeds your defined RTO, you can identify bottlenecks in the recovery process and adjust your strategy accordingly.

Regularly testing and validating your backup and recovery procedures will allow you to ensure that your application and its data are adequately protected and that you are prepared to restore them quickly and accurately in the event of a disaster.

Automation and Monitoring

Leveraging automation tools and setting up monitoring and alerting systems can help streamline your backup and recovery process, while ensuring you stay informed about any issues that could impact data recovery.

- **Automation Tools:** Implementing automation tools can simplify and standardize the backup and recovery process. This reduces the risk of human error, ensures consistency, and saves time. Cloud providers often offer built-in automation features, such as AWS Backup, Azure Backup, or Google Cloud's Backup and Restore services, that allow you to schedule and manage backups across multiple services. Alternatively, you can use third-party tools or create custom scripts to automate the process.

- **Monitoring:** Setting up monitoring systems can help you track the status of your backups and identify potential issues. Monitoring tools can provide insights into backup success rates, data integrity, and storage usage. Cloud providers typically offer built-in monitoring services, like Amazon CloudWatch, Azure Monitor, or Google Cloud Monitoring, that can be configured to track backup metrics.

- **Alerting:** Configure alerting systems to notify you of backup failures, storage capacity issues, or other problems that could impact data recovery. This allows you to quickly address any issues and minimize the risk of data loss. You can set up alerts based on specific metrics or conditions, such as backup failures or storage usage thresholds, and receive notifications via email, SMS, or other communication channels.

For example, you could set up a monitoring system to track the success rate of your nightly backups. If a backup fails, the alerting system could immediately notify you via email or SMS, allowing you to investigate and resolve the issue before it impacts your data recovery capabilities.

Implementing automation tools and setting up monitoring and alerting systems will allow you enhance the reliability and efficiency of your backup and recovery process, while staying informed about any issues that could potentially impact data recovery.

Practical: Setting Up and Managing Cloud Databases

In this section we will guide you through the hands-on process of setting up and managing cloud databases in AWS, Microsoft Azure, and Google Cloud. With a focus on practical examples, you'll learn how to effectively leverage the features and capabilities of these cloud platforms to create and manage powerful, scalable, and reliable database solutions.

You'll gain a comprehensive understanding of the various steps involved in provisioning, configuring, and maintaining cloud databases, as well as best practices for data migration, performance tuning, and ensuring high availability. By the end of this section, you will have the skills and knowledge necessary to confidently manage cloud databases in real-world scenarios, ensuring optimal performance and reliability for your applications.

Choosing a Database Service

To select the most suitable cloud database service for your specific use case, it's important to understand the features and capabilities of different cloud database services offered by AWS, Microsoft Azure, and Google Cloud.

In AWS, you have options like Amazon RDS for relational databases, Amazon DynamoDB for NoSQL databases, and Amazon Timestream for time-series databases. For in-memory databases, you can use Amazon ElastiCache. Evaluate the performance, scalability, and compatibility of each service with your application requirements to make an informed decision.

Microsoft Azure offers Azure SQL Database for relational databases, Azure Cosmos DB for NoSQL databases, and Azure Cache for Redis for in-memory databases. Assess the integration capabilities, consistency models, and pricing structure of each service to determine the best fit for your needs.

Google Cloud provides services like Cloud SQL and Cloud Spanner for relational databases, Firestore and Bigtable for NoSQL databases, and Cloud Memorystore for in-memory databases. Consider the ease of management, global scalability, and consistency guarantees of each service before making a choice.

For example, if you need a globally distributed NoSQL database with high availability, you might choose Amazon DynamoDB in AWS, Azure Cosmos DB in Microsoft Azure, or Firestore in Google Cloud. By carefully considering the features and capabilities of each cloud provider's database services, you can make the best decision for your specific use case and requirements.

Provisioning and Configuring

Provisioning and Configuring: To effectively work with cloud databases, it's crucial to become proficient in creating, configuring, and managing cloud database instances in AWS, Microsoft Azure, and Google Cloud. This includes setting up access controls, security groups, and database parameters. Here, we will provide a more detailed, hands-on example for each cloud provider:

AWS:

1. Sign in to the AWS Management Console and navigate to the Amazon RDS service.
2. Click on "Create database" and select the database engine (e.g., MySQL, PostgreSQL).
3. Choose the use case, either "Production" or "Dev/Test", based on your requirements.
4. Configure the instance specifications, such as instance class, storage type, and storage size.

5. Set up a master username and password for database access.

6. Configure the Virtual Private Cloud (VPC) and security group settings to control network access.

7. Enable automated backups and set the backup retention period.

8. Review your settings and click "Create database" to launch the instance.

Microsoft Azure:

1. Sign in to the Azure Portal and click on "Create a resource".

2. Search for "Azure SQL" and select "Azure SQL Database" from the results.

3. Click on "Create" and fill in the required details, such as subscription, resource group, database name, and server.

4. Choose the compute and storage configuration based on your needs.

5. Configure the networking settings, including firewall rules, to control access to your database.

6. Enable or disable advanced features, such as automatic tuning or threat detection.

7. Review your settings and click "Create" to deploy the database instance.

Google Cloud:

1. Sign in to the Google Cloud Console and navigate to the Cloud SQL service.

2. Click on "Create instance" and select the desired database engine (e.g., MySQL, PostgreSQL).

3. Configure the instance settings, such as instance ID, region, zone, and machine type.
4. Set up a root password for database access.
5. Configure the storage settings, including storage type and storage capacity.
6. Set up the networking settings, including authorized networks and SSL/TLS configurations.
7. Enable or disable automatic backups and set the backup window.
8. Review your settings and click "Create" to launch the instance.

Following these hands-on steps for each cloud provider will allow you gain practical experience in provisioning and configuring cloud database instances, ensuring proper access controls, security groups, and database parameters are in place for your application.

Data Migration and Import

Here are practical, hands-on examples for migrating existing data to a cloud database or importing new data, while ensuring data integrity and minimal downtime for each cloud provider:

AWS Database Migration Service (DMS):

1. Sign in to the AWS Management Console and navigate to the DMS service.
2. Click on "Create migration" and choose the source and target databases (e.g., MySQL to Amazon RDS).
3. Configure the connection details for both the source and target databases.
4. Choose the migration type (full load, CDC, or both) and select the tables and schema to migrate.

5. Start the migration task and monitor the progress using the DMS console.

Azure Database Migration Service (DMS):

1. Sign in to the Azure Portal and click on "Create a resource".
2. Search for "Azure Database Migration Service" and click "Create" to start the creation process.
3. Fill in the required details, such as subscription, resource group, and instance name.
4. Choose the desired source and target databases (e.g., SQL Server to Azure SQL Database).
5. Configure the migration activity, including connection details, schema, and tables to migrate.
6. Start the migration activity and monitor the progress using the Azure Portal.

Google Cloud Dataflow:

1. Sign in to the Google Cloud Console and navigate to the Dataflow service.
2. Click on "Create job from template" and choose a pre-built template or create a custom one for your specific data migration or import task.
3. Fill in the required details, such as job name, region, and input/output data sources (e.g., CSV files in Cloud Storage to BigQuery).
4. Configure any additional settings, such as Dataflow workers, and click "Run job" to start the migration or import process.

5. Monitor the progress of the job using the Dataflow console and check the output data in the target database.

Following these hands-on steps for each cloud provider will allow you gain practical experience in migrating existing data or importing new data to cloud databases using best practices and tools while ensuring data integrity and minimal downtime.

Monitoring and Performance Tuning

Here are practical, hands-on examples for utilizing built-in monitoring tools to analyze database performance, diagnose issues, and optimize queries and resource allocation for improved efficiency with each cloud provider:

AWS CloudWatch and Performance Insights:

1. Sign in to the AWS Management Console and navigate to the RDS or DynamoDB service.

2. Select your database instance and click on the "Monitoring" tab.

3. Use CloudWatch metrics to observe key performance indicators such as CPU utilization, read/write latency, and consumed throughput.

4. Enable Performance Insights for RDS instances to analyze query performance and identify slow or resource-consuming queries.

5. Based on the monitoring data, optimize queries or adjust resource allocation (e.g., increase instance size or provisioned throughput) for better performance.

Azure Monitor and Query Performance Insight:

1. Sign in to the Azure Portal and navigate to your Azure SQL Database or Cosmos DB instance.

2. In the "Monitoring" section, click on "Metrics" to view key performance indicators such as DTU consumption, request rates, and storage usage.

3. Use Query Performance Insight to analyze query performance and identify long-running or resource-intensive queries for Azure SQL Database instances.

4. Based on the monitoring data, optimize queries or adjust resource allocation (e.g., change service tier, increase throughput) to improve efficiency.

Google Cloud Monitoring and Cloud SQL Query Insights:

1. Sign in to the Google Cloud Console and navigate to the Cloud Monitoring service.

2. Create a new dashboard and add charts for key performance indicators such as CPU utilization, read/ write operations, and latency for your Cloud SQL, Firestore, or Bigtable instances.

3. Enable Query Insights for Cloud SQL instances to analyze query performance and identify slow or resource-consuming queries.

4. Based on the monitoring data, optimize queries or adjust resource allocation (e.g., change instance size, increase Firestore or Bigtable throughput) for better performance.

By following these hands-on steps for each cloud provider, you'll gain practical experience in using built-in monitoring tools to analyze database performance, diagnose issues, and optimize queries and resource allocation for improved efficiency.

Scaling and High Availability

Here are practical, hands-on examples for implementing strategies to ensure the scalability and high availability of your cloud database, including replication, sharding, and automatic failover with each cloud provider:

AWS RDS, DynamoDB, and Aurora:

1. Sign in to the AWS Management Console and navigate to the RDS or DynamoDB service.
2. Enable Multi-AZ deployment for your RDS instances to create a standby replica in a different availability zone for automatic failover.
3. Set up read replicas for your RDS instances to offload read traffic and improve performance.
4. For DynamoDB, enable global tables to replicate data across multiple regions for low-latency access and disaster recovery.
5. For Amazon Aurora, create read replicas and utilize Aurora Auto Scaling to adjust the number of read replicas based on demand.

Azure SQL Database and Cosmos DB:

1. Sign in to the Azure Portal and navigate to your Azure SQL Database or Cosmos DB instance.
2. Configure geo-replication for your Azure SQL Database instance to create secondary replicas in other regions for high availability and load balancing.
3. Enable auto-failover groups to automatically switch to a secondary replica in case of a primary replica failure.
4. For Cosmos DB, configure multi-region writes and global distribution to replicate data across multiple regions for low-latency access and disaster recovery.

Google Cloud SQL and Firestore:

1. Sign in to the Google Cloud Console and navigate to the Cloud SQL service.

2. Set up regional or multi-regional instances for high availability and automatic failover in case of an outage in your primary region.

3. Configure read replicas for your Cloud SQL instances to offload read traffic and improve performance.

4. For Firestore, choose a multi-region location when creating your database to ensure data is replicated across multiple regions for high availability and low-latency access.

Following these hands-on steps for each cloud provider, will allow you gain practical experience in implementing strategies to ensure the scalability and high availability of your cloud database, including replication, sharding, and automatic failover.

Module 6 Exam Questions

1. What type of database is optimized for handling time-stamped data points?

a. Relational Database

b. NoSQL Database

c. Time-Series Database

d. In-Memory Database

2. Which database service is provided by AWS?

a. MySQL

b. Microsoft SQL Server

c. Google Cloud Firestore

d. Amazon RDS

3. What type of NoSQL database is MongoDB?

a. Document-based

b. Key-value

c. Column-family

d. Graph-based

4. Which cloud provider offers the Azure SQL Database service?

a. AWS

b. Microsoft Azure

c. Google Cloud

d. IBM Cloud

5. What is the purpose of an in-memory database?

a. Store data on disk

b. Provide low-latency access and high performance

c. Handle time-series data

d. Store data in a hierarchical structure

6. Which database service is provided by Google Cloud?

a. Amazon RDS

b. Azure SQL Database

c. Google Cloud SQL

d. IBM Db2

7. What is the primary query language used by relational databases?

a. SQL

b. NoSQL

c. JSON

d. XML

8. Which database is an example of a time-series database?

a. MongoDB

b. Microsoft SQL Server

c. InfluxDB

d. Redis

9. What does RTO stand for in disaster recovery?

a. Recovery Time Objective

b. Recovery Test Objective

c. Recovery Time Order

d. Recovery Table Objective

10. In which type of cloud storage is data stored for the necessary duration while meeting compliance requirements?

a. Archive storage

b. Cold storage

c. Hot storage

d. All of the above

11. What does RPO stand for in disaster recovery?

a. Recovery Point Objective

b. Recovery Plan Objective

c. Recovery Process Objective

d. Recovery Performance Objective

12. Which database service is provided by Microsoft Azure?

a. Amazon RDS

b. Azure SQL Database

c. Google Cloud SQL

d. IBM Db2

13. What type of backup only stores the changes since the last full backup?

a. Full

b. Incremental

c. Differential

d. Continuous

14. Which AWS service can be used for in-memory caching?

a. Amazon RDS

b. Amazon ElastiCache

c. Amazon S3

d. AWS Lambda

15. What type of database is ideal for real-time analytics and caching?

a. Relational Database

b. NoSQL Database

c. Time-Series Database

d. In-Memory Database

16. Which service is used for managing cloud databases in Google Cloud?

a. Google Cloud Storage

b. Google Cloud SQL

c. Google Cloud Functions

d. Google Cloud Firestore

17. In a disaster recovery plan, what does RTO measure?

a. The acceptable amount of data loss

b. The time it takes to restore a system after a disaster

c. The maximum acceptable downtime for a system

d. The frequency of backup tests

18. Which NoSQL database is primarily used for caching and session management?

a. MongoDB

b. Redis

c. Apache Cassandra

d. Neo4j

19. Which database service is provided by AWS for NoSQL use cases?

a. Amazon RDS

b. Azure SQL Database

c. Google Cloud SQL

d. Amazon DynamoDB

20. In the context of database backup, what does the term "full backup" mean?

a. A backup of only the changes since the last backup

b. A backup of the entire database

c. A backup of all data added since the last full backup

d. A backup of only the most

Answers:

1. c
2. d
3. a
4. b
5. b
6. c
7. a
8. c
9. a
10. d
11. a
12. b
13. b
14. b
15. d
16. b
17. b
18. b
19. d
20. b

Module 7. Cloud Monitoring and Optimization

Introduction to Cloud Monitoring and Optimisation

In this module we will explore the critical aspects of monitoring and optimizing cloud-based applications and infrastructure. As organizations increasingly adopt cloud computing, it is essential to ensure that cloud resources are efficiently utilized and deliver optimal performance. Cloud monitoring and optimization are key practices that enable organizations to achieve these objectives, maintain high availability, and minimize costs.

Cloud monitoring involves collecting, analyzing, and visualizing performance metrics from various cloud resources, such as virtual machines, storage, and network components. These metrics provide valuable insights into the health and performance of your cloud environment, enabling you to detect potential issues early, troubleshoot problems, and optimize resource utilization. In this module, we will discuss various tools and techniques used for cloud performance monitoring, including both native cloud provider tools and third-party solutions.

Optimization plays a crucial role in cloud environments, as it helps in maximizing the return on investment (ROI) by identifying and addressing inefficiencies in resource allocation and usage. By continuously monitoring and analyzing your cloud infrastructure, you can identify opportunities for optimization, such as resizing instances, leveraging auto-scaling, and implementing caching mechanisms. These optimizations help ensure that your applications run smoothly and cost-effectively.

A practical example of cloud monitoring and optimization would be an e-commerce website hosted on a cloud platform. During peak times, such as Black Friday or Cyber Monday, traffic to the website may significantly increase, leading to potential performance bottlenecks and slower response times. By implementing effective cloud monitoring and optimization strategies, the website can maintain high performance and availability during these periods, while also keeping infrastructure costs under control.

Throughout this module, we will delve into various aspects of cloud monitoring and optimization, providing you with the knowledge and skills required to effectively manage and improve your cloud-based applications and infrastructure. By the end of this module, you will be well-equipped to analyze, troubleshoot, and optimize your cloud environment, ensuring that your organization reaps the full benefits of cloud computing.

Cloud Performance Monitoring Tools and Techniques

In this section, we will explore various tools and techniques for monitoring cloud performance, helping you ensure that your applications and infrastructure are running optimally and efficiently. As organizations increasingly adopt cloud computing, it is essential to maintain a strong understanding of the health and performance of your cloud resources. Effective cloud performance monitoring enables you to identify potential issues, optimize resource utilization, and maintain high availability for your applications.

We will discuss a range of monitoring tools and techniques, including native monitoring solutions provided by major cloud service providers like AWS, Azure, and GCP, as well as third-party monitoring tools that offer cross-platform support and advanced features. By understanding the different tools available and their respective capabilities, you can choose the best monitoring solution for your specific cloud environment and requirements.

In addition to tools, we will delve into key performance metrics that are crucial for understanding the health and performance of your cloud resources, such as CPU utilization, memory usage, network throughput, and storage I/O. By monitoring these metrics, you can identify potential bottlenecks, inefficiencies, and opportunities for optimization, ensuring that your cloud environment is running smoothly and cost-effectively.

We will also explore log management and analysis techniques, which are essential for gaining insights into the behavior of your applications and infrastructure. Through effective log management, you can detect anomalies, troubleshoot issues, and maintain a comprehensive understanding of your cloud environment.

Lastly, we will discuss alerting and notification strategies, which enable you to stay informed about potential issues and deviations from established performance thresholds. By configuring and managing alerts effectively, you can ensure that you are promptly notified of any problems and can take appropriate action to resolve them.

Throughout this section, we will provide practical examples to illustrate the use of various cloud performance monitoring tools and techniques, giving you a clear understanding of how to apply these concepts in real-world scenarios. By the end of this section, you will have the knowledge and skills necessary to effectively monitor and optimize your cloud-based applications and infrastructure.

Native cloud provider monitoring tools

In this point, you will learn about the monitoring tools provided by major cloud service providers, such as Amazon CloudWatch for AWS, Azure Monitor for Microsoft Azure, and Google Cloud Monitoring for Google Cloud Platform. These tools enable you to collect, analyze, and visualize performance metrics for your cloud resources within their respective environments.

Amazon CloudWatch is a comprehensive monitoring service for AWS that allows you to track various performance metrics for resources such as EC2 instances, RDS databases, and Lambda functions. With CloudWatch, you can create custom dashboards, set up alarms, and analyze log data. For example, you can use CloudWatch to monitor the CPU utilization of your EC2 instances, which can help you identify instances that may require resizing or adjustments to their auto-scaling configurations.

Azure Monitor is a versatile monitoring solution for Microsoft Azure that enables you to collect, analyze, and visualize performance metrics for resources like virtual machines, storage accounts, and web apps. Azure Monitor offers features such as metric and log analysis, application insights, and integration with other Azure services like Azure Automation and Azure Logic Apps. As an example, you can use Azure Monitor to track the response times of your web app, helping you identify performance bottlenecks and optimize your application architecture.

Google Cloud Monitoring is a robust monitoring service for Google Cloud Platform that allows you to monitor various performance metrics for resources such as Compute Engine instances, Cloud Storage buckets, and Cloud SQL databases. Google Cloud Monitoring provides features like custom dashboards, alerting policies, and integration with Google Cloud Logging. A practical use case for Google Cloud Monitoring could be monitoring the read and write latency of your Cloud SQL databases, allowing you to detect potential performance issues and optimize your database configurations.

Utilizing these native cloud provider monitoring tools will allow you gain valuable insights into the health and performance of your cloud resources, ensuring that your applications run smoothly and cost-effectively. Understanding and leveraging these tools will enable you to maintain high-performance, resilient, and cost-effective cloud environments tailored to your specific needs.

Third-party monitoring solutions

In this section, you will discover popular third-party monitoring tools that can be integrated with multiple cloud providers, such as Datadog, New Relic, and Dynatrace. These tools offer advanced features, customizable dashboards, and cross-platform support, enabling you to monitor and analyze the performance of your cloud resources effectively.

Datadog is a powerful monitoring and analytics platform that supports various cloud providers, including AWS, Azure, and GCP. With Datadog, you can collect and analyze performance metrics, log data, and distributed traces from your cloud resources, helping you gain a comprehensive understanding of your cloud environment. For example, you can use Datadog to monitor the latency of your API requests across different cloud providers, enabling you to identify and resolve any performance issues that may arise.

New Relic is a popular application performance monitoring (APM) solution that provides deep visibility into the performance of your cloud resources and applications. New Relic supports integration with multiple cloud providers and offers features such as real-time analytics, distributed tracing, and customizable dashboards. You can use New Relic to monitor the response times of your web applications hosted on AWS, Azure, or GCP, helping you optimize your application architecture and ensure high performance for your users.

Dynatrace is an all-in-one monitoring platform that offers extensive support for various cloud providers and technologies. Dynatrace provides features such as automatic discovery of your cloud resources, real-time performance analytics, and AI-driven root cause analysis. For instance, you can use Dynatrace to monitor the performance of your microservices deployed across multiple cloud platforms, allowing you to detect any bottlenecks and optimize your service architecture.

Incorporating these third-party monitoring solutions into your cloud environment will allow you effectively monitor and analyze the performance of your cloud resources across different platforms. These tools provide you with a unified view of your cloud infrastructure, allowing you to detect and resolve issues more efficiently and maintain a high-performance, cost-effective cloud environment.

Key performance metrics

In this section, you will understand the essential performance metrics to monitor in a cloud environment, including CPU utilization, memory usage, network throughput, storage I/O, and application response times. Learning how to interpret these metrics will help you identify potential bottlenecks, inefficiencies, and areas for optimization.

- **CPU utilization:** CPU utilization is a critical metric that measures the percentage of processing power being used by your cloud resources. High CPU utilization can indicate that your resource is under heavy load, which may affect the performance of your applications. By monitoring CPU utilization, you can identify when it's time to resize your instances or adjust auto-scaling policies to maintain optimal performance.

- **Memory usage:** Memory usage is an important metric that represents the amount of system memory consumed by your cloud resources. Insufficient memory can cause performance degradation or even application crashes. Monitoring memory usage helps you identify if your resources need more memory or if there's a memory leak in your application that needs to be addressed.

- **Network throughput:** Network throughput measures the amount of data being transmitted and received by your cloud resources over a period of time. Monitoring network throughput can help you identify network-related performance bottlenecks and ensure that your cloud resources have adequate bandwidth to handle incoming and outgoing traffic.

- **Storage I/O:** Storage I/O refers to the read and write operations performed on your cloud storage resources. By monitoring storage I/O metrics, such as read and write latency or throughput, you can identify performance issues related to your storage resources and optimize your storage configurations to improve overall application performance.

- **Application response times:** Application response times measure the time it takes for your application to process and respond to a user request. Monitoring application response times can help you detect performance issues and optimize your application architecture to provide a better user experience.

For example, if you notice that your application's response times are increasing, you can use the other performance metrics to identify the root cause. High CPU utilization might indicate that your instance is under-provisioned, while increasing storage I/O latency might point to a bottleneck in your database or storage system.

Understanding and monitoring these key performance metrics will allow you proactively identify potential issues and optimize your cloud environment, ensuring that your applications run smoothly and efficiently.

Log management and analysis

In this section, you will explore the importance of log management in cloud monitoring and learn how to collect, aggregate, and analyze log data from various cloud resources using tools like Elasticsearch, Logstash, and Kibana (ELK Stack), or Splunk. By effectively managing and analyzing log data, you can gain insights into application and infrastructure behavior, detect anomalies, and troubleshoot issues more effectively.

Logs are records of events generated by your cloud resources, such as application logs, system logs, or access logs. These logs contain valuable information that can help you understand the performance and behavior of your cloud environment. Proper log management involves collecting, aggregating, and analyzing log data to identify patterns, trends, and anomalies that can be used to improve the performance and reliability of your applications.

Elasticsearch, Logstash, and Kibana (ELK Stack) is a popular open-source log management solution. Logstash collects and processes log data from various sources, Elasticsearch stores and indexes the log data for fast searching, and Kibana provides a powerful visualization interface for analyzing the data. By using the ELK Stack, you can create custom dashboards to monitor application performance, detect security threats, or troubleshoot infrastructure issues.

For example, you can use the ELK Stack to analyze the access logs of your web application hosted on AWS. By visualizing the request patterns in Kibana, you might discover an unusually high number of failed login attempts, indicating a potential security threat that needs to be addressed.

Splunk is another popular log management and analysis tool that provides powerful data processing, searching, and visualization capabilities. Splunk can ingest log data from various cloud resources, allowing you to monitor, analyze, and troubleshoot your cloud environment in real-time. With Splunk, you can create custom alerts, dashboards, and reports to gain insights into your applications and infrastructure.

As an example, you can use Splunk to monitor the system logs of your virtual machines running on Microsoft Azure. By analyzing these logs, you might identify a recurring error message that indicates a misconfiguration in your virtual machine, which can be resolved to improve the overall stability of your environment.

Incorporating log management and analysis techniques into your cloud monitoring strategy will allow you gain a deeper understanding of your cloud environment and proactively address potential issues, ensuring the smooth operation and high performance of your applications and infrastructure.

Alerting and notification strategies

In this section, you will learn how to configure and manage alerts based on performance metrics and log data, ensuring that you are promptly notified of potential issues or deviations from established thresholds. Understanding best practices for alerting will help you set appropriate thresholds, reduce alert noise, and integrate with incident management systems.

Alerts are essential for maintaining a proactive monitoring strategy, as they can notify you when specific conditions are met or exceeded, such as high CPU utilization, low available storage, or an increase in application error rates. By setting up effective alerts, you can quickly identify and resolve issues before they escalate and impact your users.

Here are some best practices for configuring and managing alerts in your cloud environment:

- **Set appropriate thresholds:** When configuring alerts, it's important to set thresholds that reflect the normal operating conditions of your cloud resources. Thresholds that are too low may trigger false alarms, while thresholds that are too high might not catch genuine issues. Consider the historical performance data of your resources and the specific requirements of your applications to determine suitable threshold values.
- **Reduce alert noise:** Too many alerts can overwhelm you and make it difficult to focus on the most critical issues. To reduce alert noise, consider setting up alerts only for the most important performance metrics and log data, and use advanced filtering or aggregation techniques to minimize false positives. Additionally, you can use alert suppression or deduplication to avoid receiving multiple notifications for the same issue.
- **Integrate with incident management systems:** Integrating your alerting system with an incident management tool, such as PagerDuty or Opsgenie, can help streamline your incident response process. These tools can automatically create and assign incidents based on alerts, escalate issues to the appropriate team

members, and provide a central platform for tracking and resolving incidents.

For example, you can configure an alert in Amazon CloudWatch to notify you when the average CPU utilization of your EC2 instances exceeds 80% over a 5-minute period. By integrating this alert with your incident management system, you can automatically create an incident and assign it to the relevant team member, who can then investigate and resolve the issue.

Implementing effective alerting and notification strategies will allow you to ensure that you are quickly informed of potential issues in your cloud environment and can take appropriate action to resolve them. This proactive approach to monitoring helps maintain the high performance and reliability of your applications and infrastructure.

Analyzing and Resolving Performance Issues

In this section, we will delve into the process of analyzing and resolving performance issues in your cloud environment. As organizations increasingly rely on cloud services for their mission-critical applications and infrastructure, maintaining high performance and reliability is essential. Identifying the root cause of performance problems and implementing effective solutions can be a challenging task, but it is crucial to ensure that your applications remain responsive and available to users.

We will discuss various techniques and strategies for diagnosing and addressing performance issues, from analyzing logs and metrics to implementing infrastructure optimizations. Throughout this section, we will provide practical examples to illustrate how these approaches can be applied to real-world scenarios, helping you gain the skills needed to maintain a high-performance cloud environment.

Understanding the principles of performance analysis and optimization will allow you to be better equipped to proactively identify potential issues, troubleshoot problems effectively, and implement solutions that improve the overall performance and reliability of your cloud resources. This knowledge will enable you to deliver a superior user experience and ensure the success of your cloud-based applications and services.

Root cause analysis techniques

In this section, you will learn various methodologies and techniques for identifying the root cause of performance issues in your cloud environment, such as log analysis, metric correlations, and tracing. Understanding how to apply these techniques effectively will enable you to diagnose and resolve problems more efficiently.

- **Log analysis:** Logs contain valuable information about the behavior of your cloud resources and can provide insights into the causes of performance issues. By examining logs from your applications, databases, and infrastructure, you can identify error messages, unusual patterns, or other indicators of problems. For example, you might find that a sudden spike in error logs correlates with a performance issue, pointing you towards a potential bug or misconfiguration in your application.
- **Metric correlations:** By analyzing the relationships between different performance metrics, you can gain a deeper understanding of the underlying issues affecting your cloud environment. For instance, if you notice that high CPU utilization correlates with slow application response times, this might indicate that your cloud resources are under-provisioned or that your application is not optimized for the available resources. Correlating metrics can help you pinpoint the specific areas that need to be addressed to improve performance.
- **Tracing:** Tracing tools, such as distributed tracing or application performance monitoring (APM) solutions, can help you visualize the flow of requests through your cloud environment and identify bottlenecks or inefficiencies. By examining trace data, you can see how long each component of your application takes to process requests, making it easier to identify performance hotspots and optimize your application architecture. For example, using a tool like Jaeger or Zipkin, you might discover that a particular microservice in your application takes significantly longer to process requests than expected, indicating a potential performance issue that needs further investigation.

Mastering these root cause analysis techniques will allow you to be better equipped to diagnose and resolve performance issues in your cloud environment. This will not only help you maintain a high-performance infrastructure but also ensure that your applications deliver a seamless and reliable user experience for your users.

Performance optimization strategies

In this section, you will explore various strategies for optimizing the performance of your cloud resources, including instance resizing, load balancing, caching, and database optimization. Understanding how to apply these strategies effectively can help you improve application responsiveness and reduce latency, ensuring a seamless user experience.

- **Instance resizing:** One of the simplest ways to optimize performance is by adjusting the size of your cloud instances based on their resource requirements. If you notice that a particular instance is consistently under high CPU or memory usage, you can resize it to a larger instance type to provide more resources and alleviate performance bottlenecks. Conversely, if an instance is underutilized, downsizing to a smaller instance type can save costs without sacrificing performance.

- **Load balancing:** Distributing incoming traffic across multiple instances can help to evenly distribute the workload and prevent individual instances from becoming overloaded. By using a load balancer, you can ensure that your application remains responsive even under high traffic loads. For example, if you're hosting a web application on AWS, you can use an Application Load Balancer (ALB) to distribute incoming HTTP/HTTPS traffic across multiple EC2 instances, improving the overall performance and availability of your application.

- **Caching:** Caching is an effective strategy for reducing latency and improving the performance of data-intensive applications. By storing frequently accessed data in a cache, you can minimize the time it takes to retrieve that data and reduce the load on your backend systems. For instance, you can use a caching solution like Redis or Amazon ElastiCache to store the results of expensive database queries, reducing the need for repeated queries and improving the responsiveness of your application.

- **Database optimization:** Optimizing your database can have a significant impact on the performance of your cloud applications. Techniques such as query optimization, indexing, and partitioning can help to reduce database latency and improve query performance. Additionally, consider using managed database services, like Amazon RDS or Azure Database, which offer built-in performance optimizations and automatic scaling to meet the demands of your application.

Implementing these performance optimization strategies will allow you enhance the responsiveness and reliability of your cloud resources, ensuring that your applications consistently deliver a high-quality user experience. As you continue to monitor and analyze your cloud environment, you will be able to identify opportunities for further optimization and drive continuous performance improvements.

Capacity planning and scaling

In this section, you will learn about the importance of capacity planning in ensuring the performance and scalability of your cloud environment. Proper capacity planning helps you anticipate resource requirements, minimize costs, and prevent performance degradation due to insufficient resources.

- **Analyzing usage patterns:** To effectively plan your capacity, you should analyze your current resource usage patterns. This involves examining historical performance metrics, such as CPU utilization, memory usage, and network throughput, to identify trends and seasonal variations in demand. By understanding these patterns, you can better predict future resource requirements and allocate resources accordingly.

- **Forecasting future demand:** Based on your analysis of usage patterns, you can develop a forecast for future resource demand. This may involve using statistical methods, such as time-series analysis or regression models, to project future resource requirements based on historical data. Accurate forecasting enables you to plan for growth and ensure that you have sufficient resources in place to handle increased demand.
- **Implementing auto-scaling policies:** Auto-scaling is a powerful feature offered by many cloud providers that allows you to dynamically adjust resource allocation based on changing needs. By implementing auto-scaling policies, you can ensure that your cloud environment automatically scales up or down in response to fluctuations in demand. For example, you can configure an AWS Auto Scaling Group to automatically add or remove EC2 instances based on the average CPU utilization of the instances, ensuring that your application remains responsive during periods of high demand while minimizing costs during periods of low demand.

Understanding the principles of capacity planning and scaling will allow you proactively manage your cloud resources to ensure optimal performance and scalability. This will not only help you deliver a consistent and reliable user experience but also enable you to adapt to changing business requirements and minimize costs associated with over- or under-provisioning resources.

Infrastructure as code and automation

In this section, you will discover how using infrastructure as code (IaC) and automation tools can help you quickly implement performance optimizations and resolve issues. IaC enables you to define and manage your cloud resources using code, making it easier to apply changes consistently and automate repetitive tasks. By leveraging IaC and automation tools, you can streamline the management of your cloud environment and improve its overall performance.

- **Terraform: Terraform** is an IaC tool that allows you to define and provision cloud infrastructure using a declarative configuration language. With Terraform, you can easily create, modify, and manage your cloud resources across multiple providers, such as AWS, Azure, and Google Cloud Platform. By using Terraform to manage your infrastructure, you can quickly implement performance optimizations, such as resizing instances or adding caching layers, and ensure that these changes are applied consistently across your environment.

- **Ansible:** Ansible is an open-source automation tool that helps you automate the configuration and management of your cloud resources. By using Ansible playbooks, you can define the desired state of your infrastructure and automatically apply changes to your cloud resources. This can be especially useful for optimizing performance, as you can quickly apply updates, such as installing performance-enhancing software or adjusting configuration settings, to all relevant resources.

- **AWS CloudFormation:** AWS CloudFormation is an IaC service provided by Amazon Web Services that allows you to define and manage AWS resources using template files. With CloudFormation, you can create and update your AWS infrastructure in a consistent and repeatable manner, simplifying the process of implementing performance optimizations and resolving issues. For example, you can use CloudFormation templates to define auto-scaling policies for your EC2 instances or to create an Amazon RDS instance with performance-enhancing options enabled.

Incorporating infrastructure as code and automation tools into your cloud management processes will allow you effectively manage your cloud environment, implement performance optimizations, and resolve issues quickly. This approach not only improves the overall performance of your infrastructure but also increases the agility of your organization, making it easier to adapt to changing requirements and deliver a better user experience.

Performance testing and monitoring

In this section, you will learn about the importance of regular performance testing and continuous monitoring in maintaining a high-performance cloud environment. By conducting performance tests and continuously monitoring your cloud resources, you can proactively identify and address performance issues, ensuring a seamless user experience.

- **Regular performance testing:** Conducting regular performance tests allows you to evaluate the responsiveness, stability, and scalability of your cloud applications under various conditions, such as high traffic loads or increasing numbers of users. By identifying performance bottlenecks and areas for improvement, you can make informed decisions about resource allocation, optimization strategies, and capacity planning.

- **Performance testing tools:** There are several performance testing tools available that can help you simulate real-world user behavior and assess the performance of your cloud applications. Some popular tools include JMeter, LoadRunner, and Gatling. These tools allow you to create custom test scenarios, generate load, and analyze the results, providing valuable insights into the performance of your applications.

  For example, you can use JMeter to create a test plan that simulates users accessing your web application, with various levels of concurrency and request rates. By running this test, you can determine how your application performs under different levels of load and identify any bottlenecks or resource constraints that may be affecting performance.

- **Continuous monitoring:** In addition to performance testing, continuous monitoring is essential for maintaining a high-performance cloud environment. By monitoring your cloud resources, you can quickly detect and respond to performance issues, ensuring that your applications remain responsive and available to users.

- **Integrating performance testing and monitoring:** To maximize the effectiveness of your performance testing and monitoring efforts, it is crucial to integrate these activities into your existing monitoring and alerting systems. By doing so, you can receive timely notifications of performance issues and proactively address them before they impact your users. For instance, you can configure alerts based on performance test results, such as response times or error rates, allowing you to quickly identify and resolve issues as they arise.

Incorporating regular performance testing and continuous monitoring into your cloud management processes will allow you to ensure that your cloud environment consistently delivers a high level of performance and reliability. This proactive approach not only helps to prevent downtime and performance degradation but also enables you to continually optimize your infrastructure to meet the evolving needs of your users.

Practical: Implementing Cloud Monitoring and Alerts

Setting up monitoring tools

In this section, you will learn the steps to configure and set up native cloud provider monitoring tools, as well as third-party monitoring solutions. This hands-on approach will help you understand the process of connecting these tools to your cloud resources and collecting performance metrics.

1. Native cloud provider monitoring tools:

   a. Amazon CloudWatch (AWS): To set up CloudWatch, start by logging into your AWS Management Console. Navigate to the CloudWatch service, and create a new dashboard. You can then add widgets to display various metrics, such as CPU usage or network throughput, for your AWS resources. To collect custom metrics, you can use the AWS SDK or CLI to put data points into CloudWatch. You can also create alarms based on metric thresholds to receive notifications when certain conditions are met.

   b. Azure Monitor (Microsoft Azure): To configure Azure Monitor, sign in to the Azure portal. Navigate to the "Monitor" service, and create a new dashboard. You can add tiles to display various metrics for your Azure resources. To collect custom metrics, use the Azure Monitor REST API or the Azure SDK to send data points. You can also create alert rules based on metric thresholds to receive notifications when specific conditions are met.

   c. Google Cloud Monitoring (Google Cloud Platform): To set up Google Cloud Monitoring, sign in to the Google Cloud Console. Navigate to the "Monitoring" service, and create a new dashboard. You can add charts to display various metrics for your GCP resources. To collect custom metrics, use the Google Cloud Monitoring API or the Google Cloud SDK to send data points. You can also create alerting policies based on metric thresholds to receive notifications when certain conditions are met.

2. Third-party monitoring solutions:
   a. Datadog: To set up Datadog, sign up for an account and follow the installation instructions for your cloud platform (AWS, Azure, or GCP). This typically involves deploying a Datadog agent to your cloud resources, which collects performance metrics and sends them to the Datadog platform. Once your data is flowing into Datadog, you can create custom dashboards to visualize your metrics and configure alerts to notify you of potential issues.
   b. New Relic: To configure New Relic, sign up for an account and follow the installation instructions for your cloud platform. This usually involves deploying a New Relic agent to your cloud resources, which collects performance metrics and sends them to the New Relic platform. After your data is flowing into New Relic, you can create custom dashboards to visualize your metrics and configure alerts to notify you of potential issues.
   c. Dynatrace: To set up Dynatrace, sign up for an account and follow the installation instructions for your cloud platform. This typically involves deploying a Dynatrace OneAgent to your cloud resources, which collects performance metrics and sends them to the Dynatrace platform. Once your data is flowing into Dynatrace, you can create custom dashboards to visualize your metrics and configure alerts to notify you of potential issues.

Following these hands-on steps will allow you set up native and third-party monitoring tools for your cloud environment, connect them to your resources, and collect performance metrics. This knowledge will empower you to effectively monitor and manage the performance of your cloud resources, ensuring a seamless user experience.

Customizing dashboards and visualizations

In this section, you'll discover how to create custom dashboards and visualizations for your monitoring tools, enabling you to display key performance metrics and trends. By tailoring these visualizations to your specific needs, you can more easily identify potential issues and analyze the performance of your cloud resources.

- **Amazon CloudWatch (AWS):** To create a custom dashboard in CloudWatch, log in to your AWS Management Console and navigate to the CloudWatch service. Click on "Dashboards" in the left navigation pane and then "Create dashboard." Give your dashboard a name and click "Create." You can now add widgets to your dashboard to display various metrics. To customize a widget, click on it and configure the settings, such as choosing the desired metric, setting the time range, and selecting the aggregation method. You can also create custom CloudWatch metric math expressions to combine multiple metrics or perform calculations.

- **Azure Monitor (Microsoft Azure):** To create a custom dashboard in Azure Monitor, sign in to the Azure portal and click on "Dashboards" in the left navigation pane. Click "New dashboard" and give your dashboard a name. You can now add tiles to your dashboard to display various metrics. To customize a tile, click on it and configure the settings, such as selecting the desired metric, setting the time range, and choosing the aggregation method. You can also use Azure Monitor Workbooks to create more advanced visualizations and reports using data from multiple sources.

- **Google Cloud Monitoring (Google Cloud Platform):** To create a custom dashboard in Google Cloud Monitoring, sign in to the Google Cloud Console and navigate to the "Monitoring" service. Click on "Dashboards" in the left navigation pane and then "Create dashboard." Give your dashboard a name and click "Create." You can now add charts to your dashboard to display various metrics. To customize a chart, click on it and configure the settings, such as selecting the desired metric, setting the time range, and choosing the aggregation method. You can also use advanced features like MQL (Monitoring Query Language) to create complex queries and visualizations.

- **Datadog:** To create a custom dashboard in Datadog, log in to your account and navigate to the "Dashboards" section. Click "New Dashboard" and give your dashboard a name. You can now add widgets to your dashboard to display various metrics. To customize a widget, click on it and configure the settings, such as selecting the desired metric, setting the time range, and choosing the aggregation method. Datadog also offers advanced features like template variables and anomaly detection to create more dynamic and insightful visualizations.

- **New Relic:** To create a custom dashboard in New Relic, log in to your account and navigate to the "Dashboards" section. Click "Create a dashboard" and give your dashboard a name. You can now add widgets to your dashboard to display various metrics. To customize a widget, click on it and configure the settings, such as selecting the desired metric, setting the time range, and choosing the aggregation method. New Relic also offers advanced features like NRQL (New Relic Query Language) to create complex queries and visualizations.

Following these hands-on steps will allow you learn how to create custom dashboards and visualizations in various monitoring tools. By tailoring these visualizations to your specific needs, you'll be better equipped to identify potential issues and analyze the performance of your cloud resources effectively.

Configuring log management and analysis

In this section, you'll walk through the process of setting up log management and analysis tools, such as the ELK Stack (Elasticsearch, Logstash, and Kibana) or Splunk, to collect, aggregate, and analyze log data from your cloud resources. You'll also learn how to create custom queries and visualizations to gain insights into application and infrastructure behavior.

ELK Stack (Elasticsearch, Logstash, and Kibana):

1. **Elasticsearch:** Begin by installing Elasticsearch, the search and analytics engine that will store your log data. Follow the official Elasticsearch installation guide for your operating system.

2. **Logstash:** Install Logstash, the data processing pipeline that ingests and forwards logs to Elasticsearch. Follow the official Logstash installation guide for your operating system.

3. **Kibana:** Install Kibana, the visualization and user interface component of the stack. Follow the official Kibana installation guide for your operating system.

4. **Configure Logstash:** Create a Logstash configuration file to specify the input sources for your logs, any necessary filters, and the output destination (Elasticsearch). Ensure Logstash is properly ingesting and forwarding logs.

5. **Kibana visualizations:** Access the Kibana interface and create custom visualizations, such as bar charts, pie charts, or line charts, to display log data insights. You can also create custom queries using the Elasticsearch Query DSL to filter and aggregate log data.

Splunk:

1. **Splunk installation:** Download and install Splunk Enterprise or Splunk Cloud, depending on your requirements. Follow the official Splunk installation guide for your operating system or cloud platform.

2. **Data ingestion:** Configure Splunk to ingest log data from your cloud resources using Splunk Add-ons, such as the AWS Add-on, Azure Add-on, or Google Cloud Platform Add-on, or by configuring log forwarding from your resources.

3. **Searching and reporting:** Use the Splunk Search & Reporting app to create custom queries, reports, and visualizations based on your log data. Use the Splunk Search Processing Language (SPL) to filter, aggregate, and analyze log data.

4. **Dashboards:** Create custom dashboards in Splunk to display relevant visualizations and insights for your cloud resources. Dashboards can be tailored to specific use cases, such as application monitoring, infrastructure monitoring, or security and compliance.

Following these hands-on steps will allow you learn how to set up log management and analysis tools like the ELK Stack or Splunk, collect and analyze log data from your cloud resources, and create custom queries and visualizations to gain valuable insights into application and infrastructure behavior. This knowledge will help you detect anomalies, troubleshoot issues, and optimize your cloud environment.

Creating and managing alerts

In this section, you'll learn how to set up alerts based on performance metrics and log data to receive timely notifications of potential issues or deviations from established thresholds. You'll also understand how to configure alerting rules, reduce alert noise, and integrate with incident management systems for efficient issue resolution.

Native cloud provider tools:

- **Amazon CloudWatch:** Navigate to the CloudWatch console in your AWS Management Console. Create a new alarm by selecting the desired metric, setting a threshold, and choosing the actions that should be taken when the alarm triggers, such as sending a notification via SNS.
- **Azure Monitor:** In the Azure portal, go to the Monitor section. Create a new alert rule by selecting the target resource, specifying the metric or log data condition, and defining the actions that should be taken when the alert triggers, such as sending an email or SMS.
- **Google Cloud Monitoring:** Access the Google Cloud Console and navigate to the Monitoring section. Create a new alerting policy by defining the conditions based on metrics or log data, and specify the notification channels, such as email, SMS, or webhook.

Third-party monitoring tools:

- **Datadog:** In the Datadog web interface, navigate to the "Monitors" section. Create a new monitor by selecting the type (e.g., metric, log, or event), specifying the alert condition, and defining the notification settings, such as sending an email or integrating with an incident management system like PagerDuty.
- **New Relic:** In the New Relic One platform, access the "Alerts & AI" section. Create a new alert policy, add conditions based on metrics or log data, and set the notification channels, such as email, webhook, or integration with an incident management system.
- **Dynatrace:** In the Dynatrace web interface, navigate to the "Alerting Profiles" section. Create a new profile, define the

conditions based on performance metrics or log data, and configure the notification settings, such as sending an email or integrating with an incident management system like ServiceNow.

Reducing alert noise:

- **Use meaningful thresholds:** Set thresholds based on historical data, performance baselines, or service level objectives (SLOs) to ensure that alerts are triggered only when there's a significant deviation from normal behavior.

- **Implement alert deduplication:** Configure your alerting system to group similar alerts, so you receive only one notification for a group of related issues, reducing noise and improving focus on critical incidents.

- **Utilize alert suppression:** Temporarily disable alerts during planned maintenance or known issues to avoid being overwhelmed by irrelevant notifications.

Integrating with incident management systems:

- **PagerDuty:** Connect your monitoring and alerting tools to PagerDuty, which can automatically create incidents, route them to the appropriate team, and escalate as needed based on your on-call schedules and escalation policies.

- **Opsgenie:** Integrate your alerting tools with Opsgenie to create and manage incidents, notify on-call teams, and provide seamless communication and collaboration during incident resolution.

Following these hands-on steps will allow you learn how to create and manage alerts based on performance metrics and log data, configure alerting rules, reduce alert noise, and integrate with incident management systems for efficient issue resolution. This knowledge will help you proactively identify and address issues in your cloud environment, ensuring optimal performance and minimizing downtime.

Hands-on exercises and best practices

In this section, you'll engage in hands-on exercises to practice implementing monitoring and alerting for a sample cloud environment. You'll also receive guidance on best practices for effectively monitoring and managing the performance of your cloud resources.

Exercise: Setting up monitoring tools

1. Choose a cloud provider (AWS, Azure, or Google Cloud) or a third-party monitoring tool (Datadog, New Relic, or Dynatrace).
2. Follow the steps from the previous section to set up the monitoring tool, connect it to your cloud resources, and collect performance metrics.
3. Observe the metrics collected and ensure they accurately represent your cloud environment.

Exercise: Customizing dashboards and visualizations

1. Create a custom dashboard in your chosen monitoring tool.
2. Add visualizations for key performance metrics, such as CPU utilization, memory usage, network throughput, and application response times.
3. Customize the visualizations to display trends, highlight potential issues, and provide insights into your cloud environment's performance.

Exercise: Configuring log management and analysis

1. Set up a log management and analysis tool, such as the ELK Stack or Splunk, and connect it to your cloud resources.

2. Create custom queries and visualizations to analyze log data and gain insights into application and infrastructure behavior.

3. Use the log analysis tool to identify and troubleshoot potential issues in your cloud environment.

Exercise: Creating and managing alerts

1. Configure alerts based on performance metrics and log data in your chosen monitoring tool.

2. Set appropriate thresholds, implement alert deduplication, and utilize alert suppression to reduce alert noise.

3. Integrate your alerting system with an incident management system, such as PagerDuty or Opsgenie, for efficient issue resolution.

Best practices for monitoring and alerting:

- **Monitor a comprehensive set of metrics:** Collect a wide range of performance metrics to ensure you have a complete view of your cloud environment's performance.

- **Use meaningful thresholds:** Set alert thresholds based on historical data, performance baselines, or service level objectives (SLOs) to receive notifications when there's a significant deviation from normal behavior.

- **Regularly review and update alerting rules:** Periodically evaluate your alerting rules to ensure they remain relevant and effective as your cloud environment evolves.

- **Implement proactive monitoring:** Use monitoring and alerting tools to proactively identify and address issues before they impact end-users or result in downtime.

Engaging in these hands-on exercises and applying the best practices will allow you gain practical experience in implementing monitoring and alerting for a cloud environment. This knowledge will help you effectively manage the performance of your cloud resources, ensuring optimal performance and minimizing downtime.

Module 7 Exam Questions

1. Which native monitoring tool is associated with Amazon Web Services (AWS)?

a. Azure Monitor

b. Google Cloud Monitoring

c. Amazon CloudWatch

d. Datadog

2. Which third-party monitoring tool offers cross-platform support for multiple cloud providers?

a. Google Cloud Monitoring

b. Amazon CloudWatch

c. Azure Monitor

d. New Relic

3. What is a key performance metric to monitor in a cloud environment?

a. Number of users

b. CPU utilization

c. Color scheme

d. Logo size

4. Which tool is part of the ELK Stack for log management and analysis?

a. Ansible

b. ElasticSearch

c. Terraform

d. LoadRunner

5. What is the primary purpose of setting up alerts in a cloud environment?

a. To schedule meetings

b. To send marketing emails

c. To receive notifications of potential issues

d. To automate software updates

6. Which technique is commonly used to identify the root cause of performance issues?

a. Guessing

b. Log analysis

c. Flipping a coin

d. Asking a psychic

7. Which strategy can be used to optimize the performance of cloud resources?

a. Deleting all data

b. Ignoring performance metrics

c. Load balancing

d. Using a single large server

8. What is the purpose of capacity planning in a cloud environment?

a. To predict weather patterns

b. To plan vacations

c. To ensure performance and scalability

d. To design office layouts

9. Which tool can be used to implement infrastructure as code (IaC)?

a. Microsoft Word

b. Google Sheets

c. Adobe Photoshop

d. Terraform

10. Which performance testing tool can be integrated into your monitoring and alerting systems?

a. Microsoft Paint

b. Notepad

c. VLC Media Player

d. JMeter

11. What type of resource can be monitored with Amazon CloudWatch?

a. Azure resources

b. Google Cloud resources

c. Physical servers

d. AWS resources

12. Which visualization tool can be used with Elasticsearch for log analysis?

a. Tableau

b. Kibana

c. Google Analytics

d. Microsoft Excel

13. What is an important factor to consider when setting alert thresholds?

a. Company size

b. Office location

c. Historical data

d. Employee names

14. Which of the following is a benefit of using infrastructure as code (IaC)?

a. Improved graphics quality

b. Faster internet connection

c. Streamlined management of cloud environments

d. Easier collaboration on documents

15. What does auto-scaling do in a cloud environment?

a. Changes the color of the user interface

b. Deletes unnecessary files

c. Adjusts resource allocation based on demand

d. Creates backups of data

16. Which cloud provider offers the Azure Monitor tool?

a. IBM Cloud

b. Microsoft Azure

c. Google Cloud Platform

d. Amazon Web Services (AWS)

17. Which component of the ELK Stack is responsible for collecting log data?

a. Elasticsearch

b. Kibana

c. Splunk

d. Logstash

18. What is the primary purpose of performance testing in a cloud environment?

a. To organize email inboxes

b. To plan office events

c. To create marketing materials

d. To identify and address performance issues

19. Which tool is commonly used for database optimization in a cloud environment?

a. JMeter

b. Ansible

c. Gatling

d. LoadRunner

20. In the context of cloud monitoring, what does CPU utilization measure?

a. The color intensity of the user interface

b. The number of users connected to the system

c. The percentage of available CPU resources being used

d. The total amount of stored data

Answers:

1. a
2. d
3. c
4. a
5. b
6. a
7. a
8. b
9. a
10. a
11. a
12. a
13. a
14. a
15. a
16. b
17. b
18. a
19. d
20. c

Module 8. Cloud Cost Management and Optimization

Introduction to Cloud Cost Management and Optimization

Welcome to Module 8: Cloud Cost Management and Optimization. As businesses continue to embrace the benefits of cloud computing, it is becoming increasingly important to effectively manage and optimize cloud expenses. Effective cost management is crucial for organizations that want to maximize the value they get from their cloud investments, while maintaining control over their IT budget.

In this module, we will explore the fundamentals of cloud pricing models, techniques for reducing cloud costs, and practical strategies for analyzing and optimizing cloud expenses. You will gain a deeper understanding of how cloud providers charge for their services, and how to strategically plan and manage your cloud resources to minimize costs without sacrificing performance, reliability, or security.

Throughout this module, we will share practical examples and real-world case studies to help you better understand the different approaches to cloud cost management and optimization. You will learn about various cost-saving techniques, such as rightsizing, spot instances, and reserved instances, as well as how to effectively monitor and analyze your cloud spending to identify areas for improvement.

By the end of this module, you will be well-equipped to make informed decisions about your organization's cloud investments and have the knowledge and tools necessary to effectively manage and optimize your cloud expenses. This will not only help you maintain control over your IT budget, but also ensure that your organization is getting the most value from its cloud investments.

Stay tuned as we dive into the intricacies of cloud cost management and optimization, and empower you with the knowledge and skills to effectively manage your organization's cloud expenses.

Understanding Cloud Pricing Models

In this section, we will explore the different pricing models offered by three major cloud providers: Amazon Web Services (AWS), Microsoft Azure, and Google Cloud Platform (GCP). By understanding their pricing structures, you can make informed decisions about which provider and pricing model best suits your organization's needs.

Pay-as-you-go (On-Demand) Pricing

- **AWS:** Offers on-demand pricing for services like EC2, RDS, and S3. You pay for the resources you use, billed per hour or per request, depending on the service.

- **Azure:** Similar to AWS, Azure provides on-demand pricing for services such as virtual machines, storage, and databases. Prices are typically billed per minute or per request.

- **GCP:** Google Cloud also follows a pay-as-you-go model for services like Compute Engine, Cloud Storage, and BigQuery. They offer per-second billing for some services, providing finer cost granularity.

Reserved Instances (RI) and Savings Plans

- **AWS:** Reserved Instances are available for EC2, RDS, ElastiCache, and more. Savings Plans offer discounts for consistent usage across various AWS services, like Compute Savings Plans and EC2 Instance Savings Plans.

- **Azure:** Azure Reserved VM Instances offer discounts for long-term commitments. Azure also provides reservations for storage, SQL databases, and Cosmos DB.

- **GCP:** Google Cloud offers Committed Use Contracts for Compute Engine and discounts for long-term usage of BigQuery and Cloud Storage.

Spot Instances

- **AWS:** Spot Instances let you bid on unused EC2 capacity, offering discounts of up to 90%. Ideal for fault-tolerant workloads, they can be terminated when demand rises.
- **Azure:** Azure Spot VMs provide a similar service, offering unused capacity at discounted rates. Like AWS, instances can be terminated when demand increases.
- **GCP:** Google Cloud Preemptible VMs function like Spot Instances, offering short-lived instances at discounted rates. These instances can be terminated after a maximum of 24 hours or when demand increases.

Scaled Pricing and Tiered Discounts

- **AWS:** Offers volume discounts for services like S3 and EBS, with lower per-unit costs as usage increases.
- **Azure:** Provides tiered pricing for services like Blob Storage and Azure Functions, offering lower costs as resource consumption rises.
- **GCP:** Google Cloud follows a similar approach, offering tiered pricing for services like Cloud Storage and BigQuery, with reduced per-unit costs at higher usage levels.

Data Transfer and Egress Costs

- **AWS:** Charges for data transfer between regions, out of AWS to the internet, and between some services. Inbound data transfer is typically free.

- **Azure:** Similar to AWS, Azure charges for data transfer between regions and out to the internet. Inbound data transfer is also generally free.
- **GCP**: Charges for data egress between regions, to the internet, and between some services. Like AWS and Azure, inbound data transfer is usually free.

Understanding these pricing models and their differences across cloud providers is essential for making informed decisions about which services and cost structures best align with your organization's needs and budget.

Techniques for Reducing Cloud Costs

In this section we will explore various strategies and best practices to help you optimize your cloud infrastructure costs across major cloud providers like AWS, Microsoft Azure, and Google Cloud. Efficiently managing your cloud expenses can lead to significant savings and ensures that you are only paying for the resources you actually need, without compromising performance or user experience.

Throughout this section, we will dive into practical examples of how to implement these cost-saving techniques, which include rightsizing your resources, leveraging auto-scaling policies, taking advantage of Reserved Instances and Savings Plans, using Spot Instances or Preemptible VMs, and optimizing data transfer and caching.

Understanding and applying these cost optimization techniques, will allow you to be better equipped to manage your cloud infrastructure costs, ultimately allowing you to allocate resources more effectively and maintain control over your cloud spending.

Rightsizing

Rightsizing means regularly evaluating and adjusting the size and type of your cloud resources, such as virtual machines and storage, is essential to match the actual requirements of your workloads. This practice ensures that you are not over-provisioning or under-utilizing resources, which can lead to unnecessary costs. Below are examples of how you can apply rightsizing techniques in AWS, Microsoft Azure, and Google Cloud:

AWS:

1. Use Amazon EC2 Instance Types tailored to your workload's specific needs. For example, if your application requires high memory but low CPU, opt for a memory-optimized instance type (e.g., R-series) rather than a general-purpose instance type (e.g., M-series).
2. Leverage AWS Trusted Advisor or AWS Compute Optimizer to analyze your resource usage and receive recommendations on optimal instance sizes and types based on your workload's performance requirements.
3. Evaluate your Amazon RDS database instances and ensure that you have selected the appropriate instance class that aligns with your database's CPU, memory, and storage needs.

Microsoft Azure:

1. Utilize Azure Virtual Machine (VM) sizes that best fit your workload's requirements in terms of CPU, memory, and storage. For instance, choose a storage-optimized VM size (e.g., Ls-series) for workloads with high disk throughput and low CPU needs.
2. Monitor your VM performance with Azure Monitor and use its recommendations to resize your VMs accordingly, ensuring that you're not over-provisioning or under-utilizing resources.
3. Assess your Azure SQL Database tiers and ensure you're using the appropriate tier that matches your database's performance requirements, taking into consideration DTUs (Database Transaction Units) and vCores.

Google Cloud:

1. Select the appropriate Google Cloud Machine Types for your Compute Engine instances based on your workload's specific CPU, memory, and storage requirements. For example, if your workload is CPU-

intensive, choose a compute-optimized machine type (e.g., C2-series).

2. Utilize Google Cloud's Recommender API to receive recommendations on instance sizing and types, based on your workload's performance metrics.

3. Regularly evaluate the performance and resource usage of your Google Cloud SQL databases to ensure that you have selected the appropriate tier and machine type that matches your database's needs.

Rightsizing your cloud resources across AWS, Microsoft Azure, and Google Cloud, will allow you optimize your infrastructure costs, ensuring that you're only paying for the resources you need while maintaining the desired performance levels for your workloads.

Auto-scaling

Auto-scaling means implementing auto-scaling policies for your applications is crucial for automatically adjusting the number of instances based on real-time demand. This approach helps you maintain optimal performance while reducing costs by only using the resources needed during peak and off-peak times. Here's how you can apply auto-scaling techniques in AWS, Microsoft Azure, and Google Cloud:

AWS:

1. Utilize AWS Auto Scaling to create scaling policies for your Amazon EC2 instances. Define scaling policies based on CloudWatch metrics, such as CPU utilization or network traffic, to automatically adjust the number of instances according to demand.

2. Implement auto-scaling for your Amazon RDS instances by enabling Aurora Auto Scaling. This feature scales the

number of Aurora replicas based on the defined performance metrics and thresholds.

3. Use AWS Application Auto Scaling to manage scaling for other AWS services, like Amazon DynamoDB, Amazon ECS, and AWS Lambda, ensuring that you are only using the required resources during peak and off-peak times.

Microsoft Azure:

1. Implement Azure Virtual Machine Scale Sets to automatically scale the number of VM instances based on performance metrics and predefined rules. For example, you can create a rule to increase the number of instances when CPU usage is above 80% and decrease it when CPU usage is below 30%.

2. Leverage Azure App Service Auto Scaling for your web apps, mobile app backends, and RESTful APIs. Configure auto-scaling rules based on metrics like CPU usage, memory usage, or request rates.

3. Enable auto-scaling for Azure SQL Database by using the built-in Elastic Pool feature, which automatically scales resources based on the demand of the databases within the pool.

Google Cloud:

1. Use Google Cloud's Autoscaler for your Compute Engine instances to automatically adjust the number of instances based on real-time demand. Define autoscaling policies based on metrics like CPU utilization, network usage, or custom Stackdriver Monitoring metrics.

2. Implement auto-scaling for your Google Kubernetes Engine (GKE) clusters by configuring the Cluster Autoscaler, which adjusts the size of your cluster based on the resource requirements of your workloads.

3. Leverage App Engine's automatic scaling feature for your applications, which automatically scales the number of instances based on factors like request rate, response latency, and memory usage.

Implementing auto-scaling policies across AWS, Microsoft Azure, and Google Cloud, will allow you manage the number of instances for your applications, ensuring that you only pay for the resources you need while maintaining optimal performance during peak and off-peak times.

Reserved Instances and Savings Plans

Reserved Instances and Savings Plans: Purchasing Reserved Instances or committing to Savings Plans for predictable workloads can offer significant discounts compared to on-demand pricing. While this approach requires accurate capacity planning and forecasting, it can lead to substantial cost savings. Here's how you can take advantage of Reserved Instances and Savings Plans in AWS, Microsoft Azure, and Google Cloud:

AWS:

1. Purchase Amazon EC2 Reserved Instances for your long-running instances. Choose between Standard, Convertible, or Scheduled Reserved Instances based on your workload requirements and flexibility needs. These options offer discounted rates compared to on-demand instances, with savings up to 72%.

2. Commit to AWS Savings Plans, which provide discounts for consistent usage across various AWS services. Choose between Compute Savings Plans, which apply to EC2, Fargate, and Lambda usage, or EC2 Instance Savings Plans, which are specific to EC2 instances.

3. Consider purchasing Reserved Instances for other AWS services like Amazon RDS, Amazon ElastiCache, or

Amazon Redshift, where you can also benefit from discounted rates compared to on-demand pricing.

Microsoft Azure:

1. Buy Azure Reserved Virtual Machine Instances for your long-term, predictable workloads. Choose between 1-year or 3-year terms and benefit from discounts up to 72% compared to pay-as-you-go pricing.

2. Make reservations for other Azure services, such as Azure Blob Storage, Azure SQL Database, and Azure Cosmos DB, to secure capacity and achieve cost savings.

3. Utilize the Azure Cost Management and Billing tools to help with capacity planning and forecasting, ensuring that you make accurate reservations based on your usage patterns.

Google Cloud:

1. Commit to Google Cloud Committed Use Contracts for your long-running Compute Engine instances. Choose between 1-year or 3-year terms and benefit from discounts up to 57% compared to on-demand pricing.

2. Take advantage of committed use discounts for other Google Cloud services, like BigQuery and Cloud Storage, by committing to long-term usage.

3. Use Google Cloud's cost management tools, like the Google Cloud Pricing Calculator and Cost Reports, to assist with capacity planning and forecasting, ensuring that your committed use contracts align with your actual usage patterns.

Leveraging Reserved Instances and Savings Plans across AWS, Microsoft Azure, and Google Cloud, will allow you secure substantial cost savings for your predictable workloads while ensuring that you have the required capacity for your long-term resource needs.

Use Spot Instances or Preemptible VMs

Use Spot Instances or Preemptible VMs: Taking advantage of spot instances (AWS), spot VMs (Azure), or preemptible VMs (GCP) for fault-tolerant and flexible workloads that can handle interruptions can result in substantial cost savings. These instances offer heavily discounted rates, as they utilize unused cloud capacity, but can be terminated when demand rises. Here's how you can use Spot Instances or Preemptible VMs in AWS, Microsoft Azure, and Google Cloud:

AWS:

1. Request Amazon EC2 Spot Instances for your fault-tolerant and flexible workloads, like batch processing, big data analytics, or testing environments. Spot Instances can offer discounts up to 90% compared to on-demand pricing.
2. Use Spot Fleet to manage a collection of Spot Instances and optionally On-Demand Instances. Define your desired capacity and cost constraints, and Spot Fleet will automatically launch the optimal mix of instances to meet your requirements.
3. Implement AWS services like Spot Instance Advisor and EC2 Fleet to help you identify the best instance types and capacity pools for your Spot Instances, minimizing the risk of interruptions.

Microsoft Azure:

1. Deploy Azure Spot Virtual Machines for your workloads that can tolerate interruptions, such as development and test environments, batch processing, or stateless

applications. Azure Spot VMs offer significant cost savings compared to pay-as-you-go pricing.

2. Use Azure Virtual Machine Scale Sets with Spot VMs to scale your applications based on demand while taking advantage of the discounted rates offered by Spot VMs.

3. Leverage Azure Cost Management and Billing tools to analyze and monitor your Spot VM usage and spending, ensuring that you are optimizing costs while managing the risk of interruptions.

Google Cloud:

1. Request Google Cloud Preemptible VMs for your Compute Engine instances when running fault-tolerant and interruption-tolerant workloads, such as data processing, batch jobs, or CI/CD pipelines. Preemptible VMs can offer discounts up to 80% compared to on-demand pricing.

2. Utilize Google Cloud's managed instance groups with preemptible VMs to automatically manage and scale your instances based on demand while taking advantage of the cost savings offered by preemptible VMs.

3. Monitor your preemptible VM usage and costs with Google Cloud's cost management tools, such as the Pricing Calculator and Cost Reports, ensuring that you are effectively balancing cost optimization with the risk of interruptions.

Using Spot Instances or Preemptible VMs across AWS, Microsoft Azure, and Google Cloud, will allow you achieve significant cost savings for your fault-tolerant and flexible workloads while effectively managing the risk of interruptions due to fluctuating cloud capacity.

Optimization

Data Transfer and Caching Optimization: Minimizing data transfer and egress costs is crucial for cost-effective cloud operations. By using caching solutions (e.g., Content Delivery Networks or CDNs) to serve data from locations closer to users, you can reduce latency and data transfer costs. Additionally, optimizing data transfer between cloud services and regions helps avoid unnecessary data transfer charges. Here's how you can implement data transfer and caching optimization in AWS, Microsoft Azure, and Google Cloud:

AWS:

1. Utilize Amazon CloudFront as a CDN to cache and serve static and dynamic content, such as images, videos, or API responses, from edge locations closer to your users, reducing latency and data transfer costs.

2. Optimize data transfer between AWS services, like Amazon S3 and Amazon EC2, by selecting the same region for both services, avoiding inter-region data transfer charges.

3. Use Amazon S3 Transfer Acceleration to improve the speed of transferring data into and out of your Amazon S3 buckets, especially when transferring data across continents.

Microsoft Azure:

1. Implement Azure Content Delivery Network (CDN) to cache and serve content from Azure Blob Storage, Azure Web Apps, or custom origins at edge locations closer to your users, improving performance and minimizing data transfer costs.

2. Ensure that your Azure services, such as Azure Storage and Azure Virtual Machines, are deployed within the

same region to avoid additional data transfer fees between regions.

3. Leverage Azure Data Box or Azure ExpressRoute for large-scale data transfers to or from Azure, ensuring faster and more cost-effective data transfer compared to public internet transfers.

Google Cloud:

1. Use Google Cloud CDN to cache and serve content from Google Cloud Storage, Google Compute Engine, or custom origins at edge locations closer to your users, reducing latency and minimizing data transfer costs.

2. Deploy your Google Cloud services, like Google Cloud Storage and Google Compute Engine instances, within the same region to minimize data transfer charges between regions.

3. Opt for Google Cloud's Storage Transfer Service or Transfer Appliance for large-scale data transfers to or from Google Cloud Storage, ensuring faster and more cost-effective data transfers.

Implementing data transfer and caching optimization strategies across AWS, Microsoft Azure, and Google Cloud, will allow you minimize data transfer and egress costs while enhancing the performance and user experience of your cloud-based applications and services.

Practical: Analyzing and Optimizing Cloud Expenses

In this section we will guide you through hands-on exercises to help you gain a deeper understanding of how to analyze and optimize your cloud costs across AWS, Microsoft Azure, and Google Cloud. Gaining practical experience with cost management and optimization techniques is essential for maintaining control over your cloud spending while ensuring that your infrastructure continues to perform effectively.

During this section, you will explore real-world examples and scenarios that demonstrate how to use various cloud-native cost management tools, apply resource tagging and organization, rightsize your cloud resources, identify and eliminate unused or orphaned resources, and analyze and optimize costs in a case study.

By working through these practical exercises, you will develop the skills and knowledge necessary to effectively manage and optimize your cloud expenses, allowing you to make informed decisions about resource allocation and maintain a cost-efficient cloud infrastructure.

Utilizing Cost Management Tools

Utilizing Cost Management Tools: By leveraging native cost management tools provided by AWS, Microsoft Azure, and Google Cloud, you can effectively analyze your cloud spending patterns, identify cost-saving opportunities, and set budgets or alerts to monitor your expenses. Here's how you can use these tools in each of the major cloud providers:

AWS:

- Use AWS Cost Explorer to visualize and analyze your AWS spending, identify trends, and detect potential cost-saving

cost optimization opportunities for specific projects or departments.

Google Cloud:

- Implement labels in Google Cloud to categorize your resources, like Compute Engine instances or Cloud Storage buckets, by assigning key-value pairs that represent metadata. Use labels to organize your resources based on projects, departments, or any other custom criteria that suits your needs.
- Use Google Cloud Cost Management tools, such as the Cost Table and Cost Reports, to filter and analyze costs based on your resource labels. This allows you to gain insights into your spending patterns and uncover cost-saving opportunities for specific projects or departments.

Implementing tagging and resource organization across AWS, Microsoft Azure, and Google Cloud will allow you categorize and group your cloud resources, making it easier to analyze costs, monitor spending patterns, and identify opportunities for cost optimization.

Rightsizing and Performance Monitoring

Rightsizing and Performance Monitoring: Applying the process of evaluating and adjusting your cloud resources to match actual requirements is essential for cost optimization. By using performance monitoring tools, you can identify over-provisioned or under-utilized resources and adjust them accordingly. Here's how you can implement rightsizing and performance monitoring in AWS, Microsoft Azure, and Google Cloud:

AWS:

- Utilize Amazon CloudWatch to monitor the performance of your AWS resources, such as EC2 instances or RDS databases. Set up custom metrics and alarms to detect over-provisioned or under-utilized resources based on CPU, memory, or network usage.

- Leverage AWS Compute Optimizer to receive rightsizing recommendations for your EC2 instances and Auto Scaling groups. Use the provided insights to adjust your resources accordingly and optimize costs without compromising performance.

Microsoft Azure:

- Use Azure Monitor to track the performance of your Azure resources, like virtual machines or App Services. Configure custom metrics and alerts to identify over-provisioned or under-utilized resources based on CPU, memory, or network usage.
- Take advantage of Azure Advisor's performance recommendations to identify resources that can be rightsized. Apply the suggested changes to optimize your resource utilization and minimize costs while maintaining performance.

Google Cloud:

- Implement Google Cloud Monitoring to track the performance of your Google Cloud resources, such as Compute Engine instances or Cloud SQL databases. Set up custom metrics and alerts to detect over-provisioned or under-utilized resources based on CPU, memory, or network usage.
- Use Google Cloud Recommender to receive rightsizing recommendations for your Compute Engine instances and managed instance groups. Apply the recommended adjustments to optimize your resource utilization and reduce costs without sacrificing performance.

Incorporating rightsizing and performance monitoring practices in AWS, Microsoft Azure, and Google Cloud, will allow you to ensure that your cloud resources are optimally utilized, helping you reduce costs while maintaining the performance and reliability of your applications and services.

Identifying and Eliminating Unused or Orphaned Resources

Identifying and Eliminating Unused or Orphaned Resources: Detecting and eliminating unused or orphaned resources, such as unattached storage volumes or idle virtual machines, is crucial to avoid unnecessary cloud costs. Here's how you can identify and eliminate these resources in AWS, Microsoft Azure, and Google Cloud:

AWS:

- Use AWS Trusted Advisor to check for idle or underutilized resources, like EC2 instances or RDS databases. Review the recommendations and terminate or downsize resources that are not being utilized effectively.
- Identify unattached Elastic Block Store (EBS) volumes using the AWS Management Console, AWS CLI, or SDKs. Remove the unattached volumes to reduce costs associated with unused storage.

Microsoft Azure:

- Leverage Azure Advisor's cost recommendations to find idle or underutilized resources, such as virtual machines or SQL databases. Review the suggestions and delete or resize resources that are not being used efficiently.
- Locate unattached managed disks using the Azure Portal, Azure CLI, or SDKs. Delete the unattached disks to minimize costs related to unused storage.

Google Cloud:

- Utilize Google Cloud Recommender to discover idle or underutilized resources, like Compute Engine instances or Cloud SQL databases. Evaluate the recommendations and delete or resize resources that are not being used effectively.

- Find unattached persistent disks using the Google Cloud Console, Google Cloud CLI, or SDKs. Remove the unattached disks to cut down on costs associated with unused storage.

Actively identifying and eliminating unused or orphaned resources across AWS, Microsoft Azure, and Google Cloud, will allow you significantly reduce unnecessary cloud costs and maintain a more efficient and cost-effective infrastructure. Regularly reviewing your cloud resources and cleaning up unused or orphaned items ensures that you're only paying for the resources you truly need.

Cost Optimization Case Study

Cost Optimization Case Study: Analyzing a real-world scenario of a cloud infrastructure helps you understand how to identify areas for cost optimization and apply the techniques discussed in this module to reduce cloud expenses while maintaining performance and reliability. Here's an example of a cost optimization case study across AWS, Microsoft Azure, and Google Cloud:

Scenario: A company has a multi-cloud infrastructure with various applications, databases, and storage solutions hosted across AWS, Microsoft Azure, and Google Cloud. The goal is to optimize costs while maintaining performance and reliability.

Step 1: Analyze the current infrastructure

- Review the cloud infrastructure for each provider and document the resources, such as virtual machines, databases, and storage solutions.
- Use native cost management tools like AWS Cost Explorer, Azure Cost Management and Billing, and Google Cloud Cost Management to analyze the current spending patterns.

Step 2: Identify cost optimization opportunities

- Investigate resource utilization and performance using monitoring tools like Amazon CloudWatch, Azure Monitor, and Google Cloud Monitoring.
- Look for unused or orphaned resources, such as unattached storage volumes or idle virtual machines.
- Examine resource tagging and organization to identify any gaps in categorization or grouping.

Step 3: Implement cost optimization techniques

- Rightsize resources by adjusting their size and type to match actual requirements.
- Implement auto-scaling policies to automatically adjust the number of instances based on real-time demand.
- Purchase Reserved Instances, commit to Savings Plans, or use Spot Instances/Preemptible VMs for predictable or fault-tolerant workloads.
- Optimize data transfer and caching by using Content Delivery Networks (CDNs) and minimizing data transfer between services and regions.

Step 4: Monitor and iterate

- Continuously monitor the infrastructure using native cost management tools and performance monitoring solutions.

- Regularly review and adjust resources, auto-scaling policies, and reserved instances to ensure ongoing cost optimization.

Working through this real-world case study will allow you gain valuable insights into how to analyze cloud infrastructures, identify cost optimization opportunities, and apply the techniques discussed in this module to reduce cloud expenses while maintaining performance and reliability.

Module 8 Exam Questions

1. Which cloud provider offers Spot Instances for fault-tolerant and flexible workloads?

a) AWS

b) Microsoft Azure

c) Google Cloud

d) None of the above

2. Which tool can be used to monitor the performance of AWS resources?

a) Azure Monitor

b) Google Cloud Monitoring

c) Amazon CloudWatch

d) AWS Budgets

3. Which of the following is a cost optimization technique in the cloud?

a) Rightsizing

b) Oversizing

c) Always using on-demand instances

d) None of the above

4. What is the main purpose of using resource tagging and organization features?

a) To group and categorize cloud resources

b) To increase cloud resource performance

c) To secure cloud resources

d) To deploy cloud resources faster

5. What is the primary goal of rightsizing in the context of cloud cost optimization?

a) To match the size and type of cloud resources with actual requirements

b) To always use the largest possible resource size

c) To always use the smallest possible resource size

d) None of the above

6. Which of the following tools is used for cost analysis in Google Cloud?

a) AWS Cost Explorer

b) Azure Cost Management and Billing

c) Google Cloud Cost Management

d) Google Cloud Storage

7. Which of the following is a benefit of using auto-scaling in the cloud?

a) Reduced costs by using resources only when needed

b) Increased security

c) Decreased performance

d) None of the above

8. What is the primary purpose of using a Content Delivery Network (CDN)?

a) To serve data from locations closer to users

b) To store sensitive data

c) To increase the number of resources

d) None of the above

9. Which of the following is an example of an unused or orphaned resource?

a) Unattached storage volume

b) Virtual machine running a critical application

c) Database being actively used

d) None of the above

10. Which cloud provider offers Preemptible VMs for fault-tolerant and flexible workloads?

a) AWS

b) Microsoft Azure

c) Google Cloud

d) None of the above

11. What is the primary benefit of purchasing Reserved Instances or committing to Savings Plans?

a) Increased performance

b) Significant discounts compared to on-demand pricing

c) Unlimited resource availability

d) None of the above

12. Which of the following tools is used for cost analysis in Microsoft Azure?

a) AWS Cost Explorer

b) Azure Cost Management and Billing

c) Google Cloud Cost Management

d) Azure Resource Manager

13. Which cloud provider offers Spot VMs for fault-tolerant and flexible workloads?

a) AWS

b) Microsoft Azure

c) Google Cloud

d) None of the above

14. In the context of cloud cost optimization, what is the purpose of using performance monitoring tools?

a) To identify over-provisioned or under-utilized resources

b) To deploy new resources

c) To secure cloud resources

d) None of the above

15. Which of the following tools is used for cost analysis in AWS?

a) AWS Cost Explorer

b) Azure Cost Management and Billing

c) Google Cloud Cost Management

d) AWS CloudFormation

16. What type of workload is best suited for Spot Instances or Preemptible VMs?

a) Fault-tolerant and flexible

b) Mission-critical and inflexible

c) Always running at full capacity

d) None of the above

17. Which tool can be used to monitor the performance of Google Cloud resources?

a) Azure Monitor

b) Google Cloud Monitoring

c) Amazon CloudWatch

d) Google Cloud Storage

18. Which of the following is NOT a technique for reducing cloud costs?

a) Rightsizing

b) Auto-scaling

c) Always using on-demand instances

d) Reserved Instances and Savings Plans

19. What is the main advantage of using Spot Instances, Spot VMs, or Preemptible VMs?

a) Increased performance

b) Heavily discounted rates

c) Higher reliability

d) None of the above

20. In a multi-cloud infrastructure, which native cost management tool is used for analyzing spending patterns in AWS?

a) AWS Cost Explorer

b) Azure Cost Management and Billing

c) Google Cloud Cost Management

d) AWS Trusted Advisor

Answers:

1. a
2. c
3. a
4. a
5. a
6. c
7. a
8. a
9. a
10. c
11. b
12. b
13. b
14. a
15. a
16. a
17. b
18. c
19. b
20. a

Module 9. Cloud DevOps and Continuous Integration/ Continuous Deployment (CI/CD)

Introduction to CI/CD

Welcome to Module 9, where we will explore the world of Cloud DevOps and Continuous Integration/Continuous Deployment (CI/CD). In this module, we will discuss the significance of DevOps in modern cloud-based environments, explain the key principles and practices that drive effective CI/CD, and provide practical examples to help you understand the implementation of these concepts.

DevOps is a set of practices that bridge the gap between software development (Dev) and IT operations (Ops) teams, enabling them to work together more efficiently and effectively. In a cloud-based environment, DevOps helps organizations to accelerate the development, testing, and deployment of applications by automating various tasks and leveraging the flexibility and scalability of cloud services.

Continuous Integration (CI) is a development practice that involves developers merging their code changes into a central repository, typically multiple times a day. This process helps to catch integration issues early, reducing the time and effort required to resolve conflicts and bugs. Continuous Deployment (CD) is the practice of automatically deploying new releases to production environments once they pass all necessary tests and checks. This approach minimizes the time between development and deployment, enabling teams to deliver new features and improvements more rapidly.

CI/CD pipelines are at the heart of successful DevOps practices in cloud environments. These pipelines facilitate the automation of various stages in the software development process, such as building, testing, and deploying code. By automating these processes, teams can reduce manual intervention, minimize human error, and ensure that their applications are always in a releasable state.

In this module, we will delve into the key concepts and tools that support cloud-based DevOps processes, such as Jenkins, GitLab, and CircleCI. We will also provide hands-on examples and practical guidance on how to set up and use a CI/CD pipeline in the cloud, ensuring you have a solid foundation to build upon as you continue to explore the world of Cloud DevOps and CI/CD.

By the end of this module, you will have a comprehensive understanding of the role of DevOps in cloud-based environments, the principles and practices that underpin effective CI/CD pipelines, and the tools and techniques required to implement these concepts in real-world scenarios.

Implementing Cloud-Based DevOps Processes

In this section, we will explore the essential steps and best practices for implementing cloud-based DevOps processes. The effective implementation of DevOps practices in a cloud environment can significantly enhance collaboration between development and operations teams, streamline workflows, and enable faster and more reliable application delivery.

Cloud-based DevOps processes encompass a wide range of practices, from infrastructure management and monitoring to automated testing and deployment. These processes leverage the unique capabilities of cloud computing, such as scalability, flexibility, and on-demand resource provisioning, to optimize development and operational workflows.

In order to successfully implement cloud-based DevOps processes, it is important to understand the specific challenges and opportunities presented by cloud computing, and to identify the right tools and techniques to address them. This requires a holistic approach that considers both the technical and cultural aspects of DevOps, as well as the integration of various cloud-native services and platforms.

Throughout this section, we will provide detailed information and practical examples to help you gain a deeper understanding of the key principles and practices that drive effective cloud-based DevOps processes. We will also discuss the benefits and potential pitfalls of implementing DevOps in a cloud environment, and offer guidance on how to navigate these challenges in order to maximize the value of your DevOps initiatives.

By the end of this section, you will have a solid foundation in cloud-based DevOps processes, and will be well-equipped to begin implementing these practices in your own organization or project.

Infrastructure as Code (IaC)

Infrastructure as Code (IaC) is a key practice in cloud-based DevOps processes that allows you to manage and provision cloud resources through code, ensuring consistency and repeatability. With IaC, you can define your infrastructure in code files, which can be version-controlled, tested, and deployed just like application code. This streamlines the DevOps workflow and reduces the risk of errors and inconsistencies caused by manual processes.

Leveraging IaC will allow you easily create, update, and manage your infrastructure across various cloud platforms, such as AWS, Microsoft Azure, and Google Cloud. Here are some practical examples of how IaC can be applied to these platforms:

AWS:

In AWS, you can use AWS CloudFormation to define your infrastructure as code using JSON or YAML templates. These templates describe the resources, configurations, and dependencies required for your infrastructure, and can be version-controlled and deployed using AWS Management Console, AWS CLI, or SDKs. With CloudFormation, you can automate the provisioning and management of resources like EC2 instances, VPCs, and RDS databases, ensuring a consistent and repeatable infrastructure setup.

Microsoft Azure:

For Microsoft Azure, you can utilize Azure Resource Manager (ARM) templates to define your infrastructure as code. ARM templates are JSON files that describe the resources and configurations needed for your Azure environment. You can deploy these templates using Azure Portal, Azure CLI, or PowerShell. By using ARM templates, you can automate the creation and management of resources like Virtual Machines, App Services, and Storage Accounts, streamlining your DevOps processes.

Google Cloud:

In Google Cloud, you can employ Google Cloud Deployment Manager to manage your infrastructure as code. Deployment Manager uses YAML configuration files to define your infrastructure, along with optional Jinja2 or Python templates for parameterization and customization. You can deploy and update these configurations using Google Cloud Console, gcloud CLI, or APIs. With Deployment Manager, you can automate the provisioning and management of resources like Compute Engine instances, Cloud Storage buckets, and Cloud SQL instances, ensuring a consistent and scalable infrastructure.

In summary, Infrastructure as Code is a powerful DevOps practice that enables you to efficiently manage and provision cloud resources across various platforms. By utilizing IaC in your cloud environments, you can ensure consistency, repeatability, and streamlined workflows, ultimately improving the overall efficiency and reliability of your applications.

Automated Provisioning and Scaling

Automated provisioning and scaling are essential components of cloud-based DevOps processes. By leveraging cloud-native services and tools, you can ensure that your applications can handle fluctuations in demand without manual intervention, improving efficiency and reducing the risk of downtime. In this section, we will explore how you can utilize automated provisioning and scaling features in AWS, Microsoft Azure, and Google Cloud.

AWS:

In AWS, you can use services like Auto Scaling Groups and AWS Elastic Beanstalk to automatically provision and scale resources.

Auto Scaling Groups allow you to define scaling policies based on metrics like CPU utilization, ensuring that your Amazon EC2 instances are automatically adjusted to meet demand. Elastic Beanstalk is a Platform as a Service (PaaS) offering that simplifies the deployment, scaling, and management of applications, automatically handling capacity provisioning, load balancing, and health monitoring.

Microsoft Azure:

Azure offers various services and tools for automated provisioning and scaling, such as Azure Virtual Machine Scale Sets and Azure App Service. Virtual Machine Scale Sets enable you to create and manage a group of identical, load-balanced VMs, automatically scaling the number of VMs based on pre-defined rules and metrics. Azure App Service is a fully managed platform for building, deploying, and scaling web applications, providing auto-scaling capabilities based on custom criteria, such as response times or resource usage.

Google Cloud:

Google Cloud provides services like Managed Instance Groups and Google App Engine for automated provisioning and scaling. Managed Instance Groups allow you to create a group of identical, load-balanced Compute Engine instances, automatically scaling the number of instances based on custom rules and metrics. Google App Engine is a fully managed serverless platform for building and deploying applications, automatically scaling instances to handle incoming requests without manual intervention.

Leveraging these cloud-native services and tools will allow you efficiently automate the provisioning and scaling of your resources, ensuring that your applications are always available and performant, even during periods of high demand. This not only improves efficiency and reduces the risk of downtime but also allows you to focus on your core business objectives, rather than spending time and effort on manual infrastructure management tasks.

Monitoring and Logging:

Effective monitoring and logging are crucial aspects of cloud-based DevOps processes, as they enable you to track application performance, infrastructure health, and security events. By implementing comprehensive monitoring and logging solutions, you can proactively identify and address issues, ensuring smooth operations and faster resolution of incidents. In this section, we will discuss the monitoring and logging services available in AWS, Microsoft Azure, and Google Cloud.

AWS:

AWS provides various monitoring and logging services, such as Amazon CloudWatch and AWS CloudTrail. Amazon CloudWatch is a monitoring service that collects and tracks metrics, logs, and events from your AWS resources and applications. It enables you to set alarms and visualize data through dashboards, helping you to quickly identify and resolve issues. AWS CloudTrail is a service that records API activity for your AWS account, providing visibility into user activity, resource changes, and security events.

Microsoft Azure:

In Microsoft Azure, you can use services like Azure Monitor and Azure Log Analytics for monitoring and logging. Azure Monitor collects and analyzes performance and usage data from your Azure resources and applications, offering visualization through dashboards and allowing you to set alerts based on predefined or custom conditions. Azure Log Analytics is a powerful log search and analysis service that ingests data from various sources, including Azure Monitor, enabling you to query and analyze logs for troubleshooting, security, and compliance purposes.

Google Cloud:

Google Cloud offers services like Google Cloud Monitoring and Google Cloud Logging for comprehensive monitoring and logging. Google Cloud Monitoring, powered by Google Stackdriver, collects metrics, events, and metadata from your Google Cloud resources and applications, providing visualization through customizable dashboards and alerting based on specific conditions. Google Cloud Logging allows you to store, search, analyze, and monitor log data from your Google Cloud resources, helping you to identify trends, troubleshoot issues, and maintain security and compliance.

Implementing these monitoring and logging solutions in your cloud environment will allow you gain valuable insights into your application performance, infrastructure health, and security events. This enables you to proactively address issues, maintain smooth operations, and ensure faster resolution of incidents, ultimately contributing to the success of your DevOps initiatives.

Automated Testing and Deployment

Automated testing and deployment are critical components of a cloud-based DevOps process, enabling you to ensure that code changes are thoroughly validated and deployed to production without manual intervention. By integrating automated testing and deployment into the CI/CD pipeline, you can reduce the likelihood of introducing errors and accelerate the release of new features and improvements. In this section, we will discuss how you can automate testing and deployment in AWS, Microsoft Azure, and Google Cloud.

AWS:

AWS offers various services to support automated testing and deployment, such as AWS CodeDeploy and AWS CodePipeline. AWS CodeDeploy is a fully managed deployment service that automates software deployments to various compute services, including Amazon EC2, AWS Fargate, and AWS Lambda. AWS CodePipeline is a continuous integration and continuous delivery service that enables you to build, test, and deploy your code changes using a variety of AWS and third-party tools.

Microsoft Azure:

In Microsoft Azure, you can use services like Azure DevOps and Azure Pipelines for automated testing and deployment. Azure DevOps is a suite of services that provides end-to-end development and DevOps capabilities, including version control, agile planning, continuous integration, and delivery. Azure Pipelines is a cloud-based service that enables you to build, test, and deploy your code changes to various platforms, including Azure, Kubernetes, and Docker.

Google Cloud:

Google Cloud provides services like Google Cloud Build and Google Cloud Deployment Manager for automated testing and deployment. Google Cloud Build is a fully managed service that provides fast, consistent, and reliable builds for your applications. It can automatically build, test, and deploy your code changes to various targets, including Google Kubernetes Engine and Google Cloud Functions. Google Cloud Deployment Manager is an infrastructure-as-code service that enables you to create, configure, and manage your Google Cloud resources using templates and scripts.

Integrating automated testing and deployment into your cloud-based DevOps process will allow you to ensure that your code changes are thoroughly validated and deployed to production without manual intervention. This not only reduces the likelihood of introducing errors but also accelerates the release of new features and improvements.

Collaboration and Communication

Effective collaboration and communication are essential for a successful cloud-based DevOps process. By fostering a culture of open communication and collaboration among developers, operations teams, and other stakeholders, you can promote the sharing of knowledge, facilitate problem-solving, and drive continuous improvement and innovation. In this section, we will discuss how you can improve collaboration and communication in AWS, Microsoft Azure, and Google Cloud.

AWS:

In AWS, you can use various tools and services to promote collaboration and communication, such as Amazon Chime, Amazon Connect, and Amazon WorkSpaces. Amazon Chime is a communications service that enables you to chat, make voice and video calls, and share content securely, both inside and outside your organization. Amazon Connect is a cloud-based contact center service that enables you to provide personalized customer experiences at scale. Amazon WorkSpaces is a managed, secure cloud desktop service that enables you to provide your users with remote access to their desktops, applications, and data.

Microsoft Azure:

In Microsoft Azure, you can use services like Microsoft Teams, Azure Boards, and Azure DevOps to improve collaboration and communication. Microsoft Teams is a chat-based workspace that enables you to collaborate on projects, share files, and hold meetings securely. Azure Boards is a service that enables you to plan, track, and discuss work across teams using agile tools, Kanban boards, and reports. Azure DevOps, as previously mentioned, is a suite of services that provides end-to-end development and DevOps capabilities, including version control, agile planning, continuous integration, and delivery.

Google Cloud:

In Google Cloud, you can use services like Google Chat, Google Meet, and Google Drive to improve collaboration and communication. Google Chat is a messaging and collaboration tool that enables you to chat in real-time with individuals and groups securely. Google Meet is a video conferencing tool that enables you to hold virtual meetings with anyone, anywhere, on any device. Google Drive is a cloud-based file storage and synchronization service that enables you to share and collaborate on documents, spreadsheets, and presentations securely.

Implementing tools and platforms that facilitate information sharing, problem-solving, and decision-making, will allow you drive continuous improvement and innovation in your cloud-based DevOps process. This helps you to improve collaboration and communication among developers, operations teams, and other stakeholders, which ultimately leads to more effective and efficient software delivery.

CI/CD Tools and Automation (Jenkins, GitLab, CircleCI)

In the era of cloud-based DevOps, Continuous Integration and Continuous Deployment (CI/CD) play a crucial role in streamlining software delivery and ensuring high-quality, reliable applications. CI/CD tools and automation enable developers and operations teams to collaborate efficiently, detect issues early, and deploy changes rapidly. In this section, we will explore some of the most popular CI/CD tools and platforms, namely Jenkins, GitLab, and CircleCI, and discuss how they can be used to automate and optimize your software development pipeline.

Jenkins, GitLab, and CircleCI are powerful tools that provide an extensive set of features for automating the various stages of the software development lifecycle. These tools help you create, test, and deploy code changes in a consistent and efficient manner, reducing the likelihood of errors and accelerating the release of new features and improvements. By understanding the capabilities of these tools and integrating them into your cloud-based DevOps processes, you can significantly enhance your software delivery workflows and boost your team's productivity.

Throughout this section, we will examine the features, benefits, and use cases of Jenkins, GitLab, and CircleCI, providing practical examples of how you can leverage these tools to streamline your software development pipeline. By the end of this section, you will have a solid understanding of the key concepts and best practices related to CI/CD tools and automation, enabling you to make informed decisions about the best solutions for your specific needs and requirements.

Jenkins

Jenkins is an open-source automation server that enables you to build, test, and deploy applications efficiently. With support for hundreds of plugins that extend its functionality, Jenkins serves as a versatile tool for continuous integration and continuous delivery in your cloud-based DevOps process.

One of the key strengths of Jenkins is its extensibility, which allows you to tailor the platform to your specific needs. By leveraging the vast ecosystem of plugins, you can integrate Jenkins with various cloud platforms, such as AWS, Microsoft Azure, and Google Cloud, to streamline your CI/CD workflows. Let's explore some practical examples of how you can utilize Jenkins in these cloud environments.

AWS:

In AWS, you can use the AWS CodeDeploy plugin to integrate Jenkins with AWS CodeDeploy, enabling you to automate the deployment of your applications to EC2 instances, Lambda functions, and more. Additionally, the Amazon EC2 plugin allows you to spin up Jenkins build agents on-demand using EC2 instances, ensuring that your build and test workloads are executed on scalable, cloud-based infrastructure.

Microsoft Azure:

In Microsoft Azure, the Azure DevOps and Jenkins integration enables you to trigger Jenkins builds from Azure DevOps, streamlining your CI/CD pipeline. You can also use the Azure VM Agents plugin to create Jenkins build agents on Azure Virtual Machines, providing a scalable and cost-effective solution for running your build and test workloads.

Google Cloud:

In Google Cloud, the Google Compute Engine plugin allows you to create Jenkins build agents on Google Compute Engine instances, taking advantage of the scalability and performance of Google Cloud's infrastructure. Furthermore, the Google Kubernetes Engine plugin enables you to deploy your applications to Kubernetes clusters running on Google Cloud, simplifying your deployment process and enhancing the resilience of your applications.

By integrating Jenkins with your preferred cloud platform, you can create a seamless, efficient, and scalable CI/CD pipeline that accelerates your software delivery and promotes collaboration between developers and operations teams. Jenkins' extensibility, combined with the power of cloud computing, ensures that you can meet the challenges of modern software development and deliver high-quality, reliable applications at an unprecedented pace.

GitLab CI/CD

GitLab CI/CD is a built-in continuous integration and continuous delivery (CI/CD) service offered by GitLab, a web-based DevOps lifecycle tool. By simplifying the process of building, testing, and deploying code changes, GitLab CI/CD automates your pipeline and integrates seamlessly with GitLab's version control and collaboration features. In this section, we will discuss how you can leverage GitLab CI/CD in AWS, Microsoft Azure, and Google Cloud environments.

AWS:

In AWS, you can use GitLab CI/CD to build and test your applications, leveraging AWS services such as Amazon S3 for storing build artifacts and AWS CodeDeploy for deploying your applications to EC2 instances, Lambda functions, or other AWS compute services. By defining your pipeline in a .gitlab-ci.yml file and using GitLab's built-in AWS integrations, you can create a streamlined CI/CD process that leverages the scalability and reliability of AWS infrastructure.

Microsoft Azure:

With GitLab CI/CD, you can also automate your CI/CD pipeline in Microsoft Azure. You can use GitLab's built-in Azure integrations to deploy your applications to Azure App Service, Azure Kubernetes Service, or other Azure compute services. By defining your pipeline in a .gitlab-ci.yml file and utilizing GitLab's native Azure support, you can create an efficient and automated software delivery process that takes full advantage of Microsoft Azure's capabilities.

Google Cloud:

GitLab CI/CD can be seamlessly integrated with Google Cloud as well. You can build and test your applications using GitLab CI/CD and then deploy them to Google Cloud services such as Google Kubernetes Engine, Google App Engine, or Google Cloud Functions. By leveraging GitLab's Google Cloud integrations and defining your pipeline in a .gitlab-ci.yml file, you can create a highly efficient and automated CI/CD process that utilizes the power and flexibility of Google Cloud's infrastructure.

In conclusion, GitLab CI/CD simplifies and automates your CI/CD pipeline, making it easier for you to build, test, and deploy your applications in various cloud environments, such as AWS, Microsoft Azure, and Google Cloud. By leveraging GitLab's built-in integrations and collaboration features, you can streamline your DevOps lifecycle and ensure that your team delivers high-quality, reliable software at a rapid pace.

CircleCI

CircleCI is a cloud-based continuous integration and continuous delivery platform that automates your software development pipeline. With support for a variety of programming languages, frameworks, and deployment environments, CircleCI provides flexibility and scalability for modern DevOps workflows. In this section, we will explore how you can leverage CircleCI in AWS, Microsoft Azure, and Google Cloud environments to streamline your CI/CD processes.

AWS:

In AWS, CircleCI can be easily integrated with various AWS services to build, test, and deploy your applications. You can utilize the AWS CLI and SDKs within your CircleCI build environment to interact with services such as Amazon S3 for storing build artifacts, AWS CodeDeploy for deploying your applications to EC2 instances or Lambda functions, and more. By configuring your CircleCI pipeline using the config.yml file and leveraging AWS integrations, you can create a powerful and automated CI/CD process that takes advantage of AWS's robust infrastructure.

Microsoft Azure:

With CircleCI, you can also automate your CI/CD pipeline in Microsoft Azure. By using the Azure CLI and SDKs within your CircleCI build environment, you can interact with Azure services such as Azure Blob Storage for storing build artifacts, Azure App Service for deploying web applications, or Azure Kubernetes Service for containerized applications. By defining your pipeline in a config.yml file and making use of CircleCI's integration with Azure, you can create an efficient and scalable software delivery process that fully utilizes Microsoft Azure's capabilities.

Google Cloud:

CircleCI seamlessly integrates with Google Cloud as well. You can build and test your applications using CircleCI and then deploy them to Google Cloud services such as Google Kubernetes Engine, Google App Engine, or Google Cloud Functions. By leveraging the Google Cloud SDK and CLI tools within your CircleCI build environment and configuring your pipeline using the config.yml file, you can create a highly efficient and automated CI/CD process that harnesses the power and flexibility of Google Cloud's infrastructure.

In summary, CircleCI is a powerful cloud-based CI/CD platform that automates your software development pipeline, making it easier for you to build, test, and deploy your applications in various cloud environments like AWS, Microsoft Azure, and Google Cloud. By leveraging CircleCI's flexibility, scalability, and integrations, you can streamline your DevOps workflows and ensure that your team delivers high-quality, reliable software at a rapid pace.

Code Repository Integration

Code repository integration is a crucial aspect of efficient CI/CD processes, as it enables seamless collaboration, version control, and automation throughout your development lifecycle. Integrating CI/CD tools with popular code repositories like GitHub and Bitbucket ensures that your team can work together effectively while maintaining a high level of software quality. In this section, we will discuss how you can integrate your CI/CD tools with GitHub and Bitbucket in AWS, Microsoft Azure, and Google Cloud environments.

AWS:

In AWS, you can integrate your CI/CD tools with GitHub and Bitbucket to automate your software development pipeline. For instance, AWS CodePipeline supports direct integration with both GitHub and Bitbucket, allowing you to trigger your pipeline whenever new code is pushed to your repository. Additionally, you can use AWS CodeBuild with GitHub and Bitbucket webhooks to automatically build and test your code whenever changes are made.

Microsoft Azure:

Microsoft Azure also supports seamless integration with GitHub and Bitbucket for your CI/CD processes. Azure DevOps, for example, allows you to connect to both GitHub and Bitbucket repositories, ensuring that your pipeline is automatically triggered when new code is committed. Additionally, Azure Pipelines supports GitHub Actions and Bitbucket Pipelines, enabling you to automate your build and deployment processes within your code repository's ecosystem.

Google Cloud:

Google Cloud provides integration options for GitHub and Bitbucket as well. Google Cloud Build, a fully managed CI/CD platform, supports integration with both GitHub and Bitbucket via Cloud Build triggers. These triggers automatically start your build and test processes whenever new code is pushed to your repository. Furthermore, you can leverage Google Cloud's Source Repositories as a Git-based code repository that can easily mirror your GitHub or Bitbucket repositories, streamlining your CI/CD pipeline.

In conclusion, integrating your CI/CD tools with popular code repositories like GitHub and Bitbucket is essential for efficient collaboration, version control, and automation throughout your development lifecycle. By leveraging seamless integration options in AWS, Microsoft Azure, and Google Cloud, you can create a robust and automated software development pipeline that enables your team to deliver high-quality, reliable software at a rapid pace.

Containerization and Orchestration

Leveraging CI/CD tools in conjunction with containerization technologies like Docker and Kubernetes facilitates consistent and scalable application deployment across various environments and platforms. Containerization and orchestration are crucial for modern software development, as they enable you to package your applications and their dependencies in a uniform way, making it easier to manage, scale, and deploy your applications. In this section, we will explore how you can integrate your CI/CD tools with Docker and Kubernetes in AWS, Microsoft Azure, and Google Cloud environments.

AWS:

In AWS, you can use the Elastic Container Registry (ECR) to store and manage your Docker container images. Integrating your CI/CD pipeline with ECR allows you to automatically build and push your images during the deployment process. Additionally, AWS offers the Elastic Kubernetes Service (EKS), a managed Kubernetes service that simplifies the deployment, scaling, and management of your containerized applications. By combining your CI/CD tools with ECR and EKS, you can create a seamless and scalable deployment pipeline that takes full advantage of containerization and orchestration.

Microsoft Azure:

Microsoft Azure also supports integration with Docker and Kubernetes for your CI/CD processes. You can use Azure Container Registry (ACR) to store and manage your Docker container images, and Azure Pipelines to automate the building and pushing of images. Furthermore, Azure Kubernetes Service (AKS) provides a managed Kubernetes environment that simplifies the deployment, scaling, and management of containerized applications. By integrating your CI/CD pipeline with ACR and AKS, you can create a consistent and scalable deployment process that leverages containerization and orchestration technologies.

Google Cloud:

Google Cloud offers robust integration options for Docker and Kubernetes as well. Google Container Registry (GCR) allows you to store and manage your Docker container images, while Google Cloud Build supports the automated building and pushing of images. Additionally, Google Kubernetes Engine (GKE) provides a fully managed Kubernetes service that simplifies the deployment, scaling, and management of containerized applications. By integrating your CI/CD tools with GCR and GKE, you can create a highly efficient and automated deployment pipeline that takes advantage of containerization and orchestration.

In summary, leveraging CI/CD tools in conjunction with containerization technologies like Docker and Kubernetes ensures consistent and scalable application deployment across various environments and platforms. By integrating your CI/CD pipeline with these technologies in AWS, Microsoft Azure, and Google Cloud environments, you can create a powerful and efficient software delivery process that enables your team to deliver high-quality, reliable applications at a rapid pace.

Practical: Setting Up and Using a CI/CD Pipeline in the Cloud

In this section, we will explore the practical aspects of setting up and using a CI/CD pipeline in the cloud. The ability to implement an efficient and automated software development pipeline is crucial for modern DevOps teams, as it allows for faster delivery of high-quality software while minimizing manual intervention and errors. By leveraging cloud-based CI/CD tools and services, you can take advantage of the flexibility, scalability, and reliability of cloud infrastructure to streamline your development and deployment processes.

Throughout this section, we will guide you through the process of selecting an appropriate CI/CD tool, integrating it with your code repository, and configuring the pipeline to automate the building, testing, and deployment of your application. Additionally, we will cover how to incorporate containerization and orchestration technologies like Docker and Kubernetes into your pipeline, as well as how to implement monitoring and logging solutions to ensure smooth operations and rapid issue resolution.

Following the practical examples and recommendations provided in this section will allow you gain valuable hands-on experience in setting up and using a cloud-based CI/CD pipeline, enabling you to enhance your team's software development processes and deliver better, more reliable applications at a rapid pace.

Selecting a CI/CD Tool

Choosing a suitable CI/CD tool for your project is an important decision, as it can significantly impact the efficiency of your software development and deployment processes. To make the right choice, you should consider your team's preferences, the programming languages you use, and your target cloud environment. In this section, we will discuss the process of selecting a CI/CD tool, focusing on Jenkins, GitLab CI/CD, and CircleCI as examples, and their compatibility with AWS, Microsoft Azure, and Google Cloud.

Jenkins:

Jenkins is an open-source automation server that supports a wide range of programming languages and integrates well with many cloud environments. It offers hundreds of plugins to extend its functionality, making it a versatile choice for various projects. If your team is familiar with Jenkins or prefers an open-source solution, this could be a suitable option.

- **AWS:** Jenkins can be easily deployed on AWS using the official Jenkins Amazon Machine Image (AMI) or a containerized version on Amazon Elastic Kubernetes Service (EKS). Integration with AWS services such as CodePipeline, CodeDeploy, and Elastic Beanstalk is possible through plugins.

- **Microsoft Azure:** Azure supports Jenkins through the Azure Jenkins VM template or by deploying it on Azure Kubernetes Service (AKS). Jenkins can be integrated with Azure DevOps and Azure Pipelines, among other Azure services.

- **Google Cloud:** Jenkins can be deployed on Google Cloud using the Jenkins Certified by Bitnami solution, or as a container in Google Kubernetes Engine (GKE). Integration with Google Cloud services, such as Cloud Build and Cloud Run, is achievable through available plugins.

GitLab CI/CD:

GitLab CI/CD is a built-in continuous integration and continuous delivery service offered by GitLab, a web-based DevOps lifecycle tool. It provides seamless integration with GitLab's version control and collaboration features, making it an ideal choice if you are already using GitLab for your project.

- **AWS:** You can deploy GitLab on AWS using the official GitLab Enterprise Edition (EE) AMI, or run GitLab CI/CD as a standalone service. Integration with AWS services, such as CodePipeline and ECR, can be achieved using GitLab Runners.

- **Microsoft Azure:** GitLab can be deployed on Azure using the GitLab EE Azure Marketplace template, or by running GitLab CI/CD as a standalone service. Integration with Azure services, such as Azure Pipelines and ACR, is possible using GitLab Runners.

- **Google Cloud:** GitLab can be deployed on Google Cloud using the GitLab EE Certified by Bitnami solution, or as a container in GKE. Integration with Google Cloud services, such as Cloud Build and GCR, can be achieved using GitLab Runners.

CircleCI:

CircleCI is a cloud-based CI/CD platform that supports a variety of programming languages, frameworks, and deployment environments. It provides flexibility and scalability for modern DevOps workflows and may be a suitable choice if you prefer a fully managed, cloud-based solution.

- **AWS:** CircleCI integrates with AWS services like CodeDeploy, ECR, and EKS using built-in orbs, which are reusable, shareable packages that streamline your pipeline configuration.

- **Microsoft Azure:** CircleCI can integrate with Azure services such as Azure DevOps, Azure Pipelines, and ACR using available orbs, making it compatible with your Azure-based projects.

- **Google Cloud:** CircleCI supports integration with Google Cloud services like Cloud Build, GCR, and GKE using orbs, allowing you to easily configure your pipeline for deployment in Google Cloud.

In conclusion, selecting a suitable CI/CD tool for your project is crucial for streamlining your development and deployment processes. Consider your team's preferences, the programming languages you use, and your target cloud environment when making a decision. By choosing a tool that aligns with your team's needs and workflow, you can greatly enhance collaboration, automate repetitive tasks, and ensure consistent and reliable delivery of your applications. Ultimately, this leads to faster releases, higher-quality software, and greater overall efficiency for your organization.

Integrating with Code Repositories

Connecting your CI/CD pipeline with your code repository, such as GitHub or Bitbucket, enables automated triggering of builds and deployments upon code changes. This seamless integration is crucial for modern software development, as it helps accelerate the development process while ensuring that your application is always up to date with the latest code changes. In this section, we will discuss how you can integrate your CI/CD pipeline with popular code repositories in AWS, Microsoft Azure, and Google Cloud environments.

AWS:

To integrate your CI/CD pipeline with a code repository in AWS, you can leverage AWS CodeCommit, a fully managed source control service that hosts Git repositories. By connecting your pipeline with CodeCommit, you enable automatic triggering of builds and deployments whenever there are code changes. Additionally, you can integrate your CI/CD tools with other popular repositories like GitHub and Bitbucket using available plugins or built-in functionality.

Microsoft Azure:

In Microsoft Azure, you can connect your CI/CD pipeline with Azure Repos, a Git-based source control service that is part of Azure DevOps. By integrating Azure Repos with your pipeline, you can automate the triggering of builds and deployments upon code changes. Furthermore, Azure Pipelines, a cloud-based CI/CD service, provides native integration with popular code repositories like GitHub and Bitbucket, allowing for seamless collaboration and automation within your development process.

Google Cloud:

Google Cloud does not offer a native code repository service like AWS or Azure. However, you can still integrate your CI/CD pipeline with popular code repositories like GitHub and Bitbucket using available plugins or built-in functionality. For example, Google Cloud Build, a fully managed CI/CD platform, provides native integration with GitHub and Bitbucket, enabling automatic triggering of builds and deployments whenever there are code changes in your repositories.

In summary, integrating your CI/CD pipeline with your code repository enables automated triggering of builds and deployments upon code changes, accelerating the development process and ensuring your application is up to date with the latest changes. By connecting your pipeline with popular code repositories like GitHub and Bitbucket, or using cloud-native source control services like AWS CodeCommit or Azure Repos, you can streamline your development and deployment processes and enhance collaboration within your team.

Setting Up Build and Test Processes

Configuring your CI/CD pipeline to automate the building and testing of your application is essential for ensuring that code changes are validated before being deployed to production. This automated process minimizes the risk of introducing errors into your application and accelerates the release of new features and improvements. In this section, we will discuss how you can set up build and test processes in your CI/CD pipeline across AWS, Microsoft Azure, and Google Cloud environments.

AWS:

In AWS, you can use AWS CodeBuild, a fully managed build service, to automate the building and testing of your application. CodeBuild supports a wide range of programming languages, frameworks, and build tools, allowing you to configure custom build environments and testing processes. By integrating CodeBuild with your CI/CD pipeline, you can automatically validate code changes and ensure they meet your quality standards before being deployed to production.

Microsoft Azure:

Azure Pipelines, a cloud-based CI/CD service, provides native support for building and testing your application in a wide variety of programming languages and frameworks. You can configure your build and test processes using YAML or a visual designer, specifying the necessary steps, tools, and dependencies required for your application. Azure Pipelines automatically validates your code changes, ensuring they meet your quality standards and are ready for deployment.

Google Cloud:

Google Cloud Build, a fully managed CI/CD platform, allows you to automate the building and testing of your application by creating custom build configurations using YAML files. Cloud Build supports a wide range of programming languages, frameworks, and build tools, enabling you to configure custom build environments and testing processes tailored to your application's requirements. By integrating Cloud Build with your CI/CD pipeline, you can automatically validate code changes and ensure they meet your quality standards before deployment.

In conclusion, setting up automated build and test processes in your CI/CD pipeline is essential for ensuring that code changes are validated before being deployed to production. By configuring your pipeline to automate the building and testing of your application in AWS, Microsoft Azure, or Google Cloud, you can minimize the risk of introducing errors and accelerate the release of new features and improvements, enhancing the efficiency and quality of your software development process.

Implementing Containerization and Orchestration

Leveraging containerization and orchestration technologies like Docker and Kubernetes in your CI/CD pipeline allows you to package your application and its dependencies in a consistent manner. This approach enables scalable and efficient deployment across different environments and platforms, improving the overall reliability and flexibility of your software development process. In this section, we will discuss how you can implement containerization and orchestration in your CI/CD pipeline across AWS, Microsoft Azure, and Google Cloud environments.

AWS:

In AWS, you can utilize Docker to containerize your application and package it with all the necessary dependencies. To automate the building of Docker images, you can use AWS CodeBuild, which supports building Docker images directly from your source code repository. For orchestration, AWS offers Amazon Elastic Kubernetes Service (EKS) and Amazon Elastic Container Service (ECS), both of which can be integrated with your CI/CD pipeline to automatically deploy and manage your containerized application.

Microsoft Azure:

Azure supports containerization using Docker, allowing you to package your application and its dependencies in a consistent manner. You can automate the building of Docker images using Azure Pipelines, which provides native support for building Docker images directly from your source code repository. For orchestration, Azure offers Azure Kubernetes Service (AKS) and

Azure Container Instances (ACI), both of which can be integrated with your CI/CD pipeline to automatically deploy and manage your containerized application.

Google Cloud:

Google Cloud enables you to containerize your application using Docker, ensuring consistent packaging of your application and its dependencies. To automate the building of Docker images, you can use Google Cloud Build, which supports building Docker images directly from your source code repository. For orchestration, Google Cloud provides Google Kubernetes Engine (GKE) and Cloud Run, both of which can be integrated with your CI/CD pipeline to automatically deploy and manage your containerized application.

In summary, implementing containerization and orchestration technologies like Docker and Kubernetes in your CI/CD pipeline enhances the efficiency and scalability of your software development process. By packaging your application and its dependencies in a consistent manner and leveraging orchestration services in AWS, Microsoft Azure, or Google Cloud, you can deploy your application seamlessly across different environments and platforms, improving the overall reliability and flexibility of your development process.

Monitoring and Logging

Incorporating monitoring and logging solutions into your CI/CD pipeline is essential for tracking application performance, infrastructure health, and security events. By doing so, you can proactively identify and resolve issues, ensuring smooth operations and faster resolution of incidents. In this section, we will discuss how you can integrate monitoring and logging solutions in your CI/CD pipeline across AWS, Microsoft Azure, and Google Cloud environments.

AWS:

In AWS, you can use Amazon CloudWatch to monitor and log your application performance and infrastructure health. CloudWatch collects and tracks metrics, logs, and events, providing you with a comprehensive view of your application and its underlying resources. By integrating CloudWatch with your CI/CD pipeline, you can automatically monitor the health of your deployed application, detect anomalies, and trigger alerts based on predefined thresholds. For security events, you can use AWS Security Hub, which aggregates and prioritizes security findings from various AWS services and partner tools.

Microsoft Azure:

Azure Monitor is Microsoft's comprehensive monitoring and logging solution that allows you to track application performance, infrastructure health, and security events. By integrating Azure Monitor with your CI/CD pipeline, you can automatically collect and analyze telemetry data from your application and its underlying resources, enabling you to identify and address issues proactively. In addition, Azure Security Center provides centralized monitoring of security events and alerts, helping you maintain a strong security posture across your application and infrastructure.

Google Cloud:

Google Cloud offers Cloud Monitoring and Cloud Logging, which together provide a comprehensive solution for monitoring and logging your application performance, infrastructure health, and security events. By integrating these services with your CI/CD pipeline, you can automatically collect and analyze telemetry data, identify anomalies, and trigger alerts based on predefined thresholds. For security event monitoring, you can use Google Cloud Security Command Center, which aggregates and prioritizes security findings from various Google Cloud services and partner tools.

In conclusion, incorporating monitoring and logging solutions into your CI/CD pipeline is crucial for tracking application performance, infrastructure health, and security events. By integrating monitoring and logging services like Amazon CloudWatch, Azure Monitor, or Google Cloud Monitoring and Logging with your pipeline, you can proactively identify and resolve issues, ensuring smooth operations and faster resolution of incidents. This ultimately enhances the reliability and efficiency of your software development process.

Module 9 Exam Questions

1. What is the primary goal of implementing Infrastructure as Code (IaC)?

a. To reduce the cost of cloud resources

b. To ensure consistency and repeatability in managing cloud resources

c. To improve communication between teams

d. To speed up the development process

2. Which AWS service can be used to automate the building and testing of your application?

a. AWS CodeCommit

b. AWS CodeBuild

c. AWS CodeDeploy

d. AWS CodePipeline

3. Which Azure service provides native support for building Docker images directly from your source code repository?

a. Azure DevOps

b. Azure Pipelines

c. Azure Container Registry

d. Azure Container Instances

4. What is the purpose of using Docker in your CI/CD pipeline?

a. To manage cloud resources

b. To package your application and its dependencies consistently

c. To deploy your application to different platforms

d. To monitor application performance

5. Which Google Cloud service can be used to build Docker images from your source code repository?

a. Google Cloud Run

b. Google Kubernetes Engine

c. Google Cloud Build

d. Google Cloud Functions

6. Which CI/CD tool is a built-in service offered by GitLab?

a. Jenkins

b. GitLab CI/CD

c. CircleCI

d. Travis CI

7. What is the primary benefit of automated provisioning and scaling in the cloud?

a. Reducing the number of developers needed

b. Handling fluctuations in demand without manual intervention

c. Increasing revenue

d. Simplifying code repositories

8. Which AWS service can be used for container orchestration?

a. Amazon ECS

b. AWS CodeBuild

c. AWS Lambda

d. AWS App Runner

9. Which Microsoft Azure service can be used for container orchestration?

a. Azure Container Instances

b. Azure Functions

c. Azure Kubernetes Service

d. Azure DevOps

10. Which Google Cloud service can be used for container orchestration?

a. Google Cloud Functions

b. Google Kubernetes Engine

c. Google Cloud Run

d. Google App Engine

11. Which AWS service can be used for monitoring application performance and infrastructure health?

a. AWS CloudTrail

b. Amazon CloudWatch

c. AWS X-Ray

d. AWS AppConfig

12. What is the primary purpose of incorporating monitoring and logging solutions into your CI/CD pipeline?

a. To track application performance, infrastructure health, and security events

b. To minimize cloud resource costs

c. To improve team communication

d. To increase the speed of software development

13. Which Azure service can be used for monitoring application performance and infrastructure health?

a. Azure Application Insights

b. Azure Log Analytics

c. Azure Monitor

d. Azure Security Center

14. Which Google Cloud service can be used for monitoring application performance and infrastructure health?

a. Google Cloud Monitoring

b. Google Cloud Logging

c. Google Cloud Trace

d. Google Cloud Debugger

15. What is the main goal of fostering collaboration and communication in the DevOps process?

a. To reduce costs

b. To speed up application development

c. To drive continuous improvement and innovation

d. To ensure security

16. Which CI/CD tool is an open-source automation server?

a. Jenkins

b. GitLab CI/CD

c. CircleCI

d. Travis CI

17. Which containerization technology is commonly used in CI/CD pipelines?

a. OpenShift

b. Kubernetes

c. Docker

d. Mesos

18. In the context of CI/CD, what does CI stand for?

a. Container Infrastructure

b. Cloud Integration

c. Continuous Integration

d. Code Inspection

19. In the context of CI/CD, what does CD stand for?

a. Cloud Deployment

b. Continuous Delivery

c. Code Deployment

d. Container Deployment

20. Which cloud provider offers the Azure Monitor service for monitoring application performance and infrastructure health?

a. Amazon Web Services

b. Microsoft Azure

c. Google Cloud Platform

d. IBM Cloud

Answers:

1. b
2. b
3. b
4. b
5. c
6. b
7. b
8. a
9. c
10. b
11. b
12. a
13. c
14. a
15. c
16. a
17. c
18. c
19. b
20. b

Module 10. Cloud Transformation Project

Introduction to the Cloud Project

Welcome to Module 10. This module focuses on the practical application of the knowledge and skills you've acquired throughout the course. By the end of this module, you will have a deeper understanding of the process of cloud transformation and the considerations that must be taken into account when transitioning from a traditional IT infrastructure to a cloud-based infrastructure.

Cloud transformation is an essential aspect of modern IT management, enabling businesses to leverage the power of the cloud to optimize performance, increase scalability, and reduce operational costs. However, successful cloud transformation projects require careful planning, strategic decision-making, and effective execution. In this module, we will explore the key components of a successful cloud transformation project, using real-world examples to illustrate the challenges and opportunities that organizations may face during this process.

One such example is the migration of a company's on-premises data center to a cloud-based infrastructure. This could involve the migration of applications, databases, and other critical systems to a cloud platform, such as Amazon Web Services (AWS), Microsoft Azure, or Google Cloud Platform (GCP). During this process, the organization must carefully evaluate its existing infrastructure, identify areas for optimization, and develop a comprehensive migration plan that accounts for potential risks and ensures minimal disruption to business operations.

Another example is the transformation of a company's software development process to adopt cloud-native technologies and practices. This might involve the adoption of containerization, microservices architecture, and serverless computing, as well as the implementation of DevOps and CI/CD pipelines to improve the efficiency and agility of the development process.

In this module, you will have the opportunity to apply the concepts and methodologies you've learned to a real-world cloud transformation use case. You will identify the critical components of a successful cloud transformation project, develop a comprehensive plan, and execute the project, taking into account the various challenges and opportunities that may arise during the process.

By the end of this module, you will have a deeper understanding of the practical aspects of cloud transformation and will be equipped with the knowledge and skills necessary to successfully plan and execute a cloud transformation project in your own organization or for your clients.

Identifying a Real-World Cloud Transformation Use Case

In this section, we will delve into the crucial process of identifying a real-world cloud transformation use case that aligns with your organization's needs and objectives. The ability to pinpoint the most suitable use case is essential for ensuring a successful cloud migration and maximizing the benefits of the transformation.

Identifying the right use case requires a comprehensive understanding of the organization's current IT infrastructure, strategic goals, and industry-specific requirements. It is important to consider not only the technical aspects of the transformation but also the potential impact on business operations, ensuring a seamless transition with minimal disruption.

Throughout this section, we will explore practical examples of cloud transformation use cases across various industries, providing insights into the different factors that influence the decision-making process. These examples will highlight the diverse range of cloud transformation scenarios, from migrating an on-premises data center to adopting cloud-native technologies for software development.

By the end of this section, you will be equipped with the knowledge and understanding necessary to identify the most suitable real-world cloud transformation use case for your organization or clients. This foundation will enable you to develop a comprehensive cloud transformation plan that addresses the unique challenges and opportunities presented by your chosen use case.

Assessing the current IT infrastructure

Assessing the current IT infrastructure is a critical first step in identifying areas that could benefit from cloud migration. By examining your existing infrastructure, including hardware, software, and network components, you can gain a clear understanding of your organization's needs and the potential benefits of cloud transformation.

To begin the assessment process, start by creating an inventory of your current IT assets. This inventory should include all physical and virtual servers, storage devices, networking equipment, and software applications. Document the specifications, configurations, and dependencies of each asset, as this information will be crucial in determining the most suitable cloud migration strategy.

Next, analyze the performance, utilization, and costs associated with each asset. This analysis will help you identify underutilized resources or bottlenecks that could be addressed through cloud migration. For example, you might find that certain servers are consistently running at low capacity, while others are struggling to handle peak demand. In this case, migrating to a cloud-based infrastructure could provide greater scalability and more efficient resource utilization.

Additionally, consider the age and lifecycle of your IT assets. Older hardware may be more prone to failure and may not support the latest technologies, making it a prime candidate for replacement with cloud-based alternatives. Similarly, legacy software applications that require extensive maintenance and support could benefit from being re-architected or replaced with cloud-native solutions.

Once you have a comprehensive understanding of your current IT infrastructure, you can identify the areas that would benefit most from cloud migration. This may include migrating compute resources to virtual machines or containers in the cloud, adopting cloud-based storage and database solutions, or leveraging cloud-based network services to improve performance and security.

Assessing your current IT infrastructure will allow you to develop a targeted cloud transformation strategy that addresses your organization's specific needs and maximizes the benefits of the migration.

Identifying the appropriate cloud platform

Identifying the appropriate cloud platform is a vital step in ensuring the success of your cloud transformation project. As you evaluate the various cloud platforms available, such as Amazon Web Services (AWS), Microsoft Azure, or Google Cloud Platform (GCP), it's essential to consider how each platform aligns with your requirements, budget, and strategic goals.

To make an informed decision, begin by comparing the features, services, and pricing models of each cloud provider. Each platform offers a unique set of services and capabilities tailored to different use cases, from compute and storage resources to advanced analytics and machine learning tools. Consider the specific requirements of your cloud transformation project and determine which platform offers the best combination of features and services to meet those needs.

Next, assess the cost implications of each platform. Cloud providers typically offer a range of pricing models, including pay-as-you-go, reserved instances, and spot instances, which can impact the overall cost of your cloud transformation project. Compare the costs of each platform, taking into account not only the direct expenses, such as compute and storage costs, but also indirect costs, such as data transfer fees and support services.

In addition to features and cost, consider other factors that may influence your decision, such as the platform's integration capabilities, security features, and compliance certifications. For example, if your organization relies heavily on Microsoft products,

Azure may offer better integration with your existing tools and services. Similarly, if you operate in a highly regulated industry, it's crucial to select a cloud provider with the necessary compliance certifications, such as HIPAA or GDPR.

Finally, examine the availability and performance of each cloud platform. Cloud providers typically offer global infrastructure with multiple regions and availability zones, which can affect the performance and reliability of your cloud-based services. Evaluate the geographic reach of each platform and select the one that best aligns with your organization's needs in terms of latency, redundancy, and disaster recovery.

Thoroughly evaluating the various cloud platforms will allow you identify the one that best aligns with your organization's requirements, budget, and strategic goals, ensuring a successful cloud transformation project.

Analyzing the organization's goals and objectives

Analyzing your organization's goals and objectives is crucial in determining the specific use cases for cloud transformation that can contribute to achieving these objectives. By understanding both short-term and long-term goals, such as cost reduction, increased scalability, improved performance, or enhanced security, you can ensure that your cloud transformation project aligns with and supports your organization's overall strategy.

Start by engaging with key stakeholders within your organization to gain a comprehensive understanding of their goals and objectives. This may include discussions with executive leadership, IT management, and business unit leaders to identify areas where cloud transformation could have the most significant impact on your organization's success.

Once you have a clear understanding of your organization's goals, consider how cloud transformation can help achieve those objectives. For example:

- **Cost reduction:** Migrating to a cloud-based infrastructure can often result in cost savings through reduced capital expenditures on hardware, lower maintenance costs, and more efficient resource utilization. Analyze your current IT spending and identify areas where cloud migration could lead to cost savings.
- **Increased scalability:** Cloud platforms offer virtually unlimited scalability, allowing your organization to quickly respond to changing business demands. Identify workloads or applications that could benefit from increased scalability and evaluate the potential benefits of migrating these to the cloud.
- **Improved performance:** Cloud providers offer a range of services designed to optimize the performance of your applications, such as content delivery networks (CDNs), load balancers, and caching services. Assess your current application performance and determine if adopting these cloud-based services could improve the user experience.
- **Enhanced security:** Cloud providers invest heavily in security infrastructure and offer a range of tools and services designed to protect your data and applications. Evaluate your organization's security posture and consider how leveraging cloud-based security services, such as encryption, identity management, and threat detection, could enhance your overall security.

Aligning your cloud transformation project with your organization's goals and objectives will allow you to ensure that your migration delivers tangible benefits and contributes to the long-term success of your organization.

Examining industry-specific requirements

Examining industry-specific requirements is a crucial aspect of the cloud transformation process, as it helps you take into account any regulations or compliance requirements that may impact your migration. Different industries have unique requirements, such as strict data privacy and security standards, which can influence the choice of cloud provider and the overall cloud transformation strategy.

To begin, research and familiarize yourself with the regulations and compliance requirements relevant to your industry. This may include standards such as the Health Insurance Portability and Accountability Act (HIPAA) for healthcare organizations, the Payment Card Industry Data Security Standard (PCI DSS) for companies handling credit card transactions, or the General Data Protection Regulation (GDPR) for organizations operating within the European Union.

Once you have a clear understanding of your industry-specific requirements, consider how they will impact your cloud transformation project:

- **Cloud provider selection:** Choose a cloud provider that offers the necessary compliance certifications and can support the required security and privacy standards. Review the provider's documentation, certifications, and third-party audit reports to ensure they can meet your industry-specific requirements.

- **Data storage and processing:** Identify where your sensitive data will be stored and processed in the cloud, and ensure that the chosen cloud platform complies with the relevant data protection regulations. This may involve selecting specific regions or availability zones, or implementing additional security measures, such as encryption or data masking.

- **Access controls and identity management:** Implement robust access controls and identity management solutions to ensure that only authorized personnel can access sensitive data and resources within the cloud environment.

This may involve using cloud-based identity and access management services or integrating with your existing on-premises solutions.

- **Monitoring and auditing:** Establish monitoring and auditing processes to track user activity, detect potential security threats, and ensure ongoing compliance with industry-specific requirements. Leverage cloud-native monitoring and logging tools, as well as third-party solutions, to maintain visibility and control over your cloud environment.

Carefully examining industry-specific requirements and incorporating them into your cloud transformation strategy, will allow you to ensure that your migration remains compliant with the necessary regulations and standards while minimizing potential risks and disruptions to your business operations.

Evaluating the potential impact on business operations

Evaluating the potential impact of cloud transformation on your daily operations is essential for ensuring a smooth transition and minimizing disruption to your business. By considering factors such as potential downtime, employee training, and changes to existing processes, you can identify the most suitable real-world use case for cloud transformation that aligns with your organization's needs.

To assess the potential impact on business operations, consider the following:

- **Potential downtime:** Determine the extent to which your cloud transformation project may cause downtime or disruptions to your services. Plan for potential outages by scheduling migration activities during periods of low usage or by adopting a phased migration approach to minimize the impact on your customers and stakeholders.

- **Employee training:** Identify the skills and knowledge required for your team to effectively manage and maintain your new cloud-based infrastructure. Develop a comprehensive training plan to ensure your employees are well-equipped to handle the transition, including hands-on workshops, online courses, or certifications from your chosen cloud provider.
- **Changes to existing processes:** Analyze how your cloud transformation project will affect existing business processes, such as software development, IT operations, or data management. Identify any necessary adjustments and develop a plan to implement these changes while minimizing disruption to your daily operations.
- **Communication and stakeholder engagement:** Engage with key stakeholders throughout the cloud transformation process to ensure they are informed of the potential impact on business operations and can provide input on critical decisions. Develop a clear communication plan to keep all relevant parties informed of progress, challenges, and any necessary adjustments to the project timeline.
- **Risk management:** Identify potential risks associated with your cloud transformation project, such as security threats, data loss, or compliance violations. Develop a comprehensive risk management plan to mitigate these risks and ensure that your migration proceeds smoothly and securely.

Thoroughly evaluating the potential impact of cloud transformation on your business operations will allow you to develop a comprehensive migration plan that minimizes disruptions, addresses potential challenges, and ensures a successful transition to a cloud-based infrastructure.

Developing a Cloud Transformation Plan

In this section we will discuss the process of creating a comprehensive plan for your organization's cloud migration. A well-structured cloud transformation plan is essential for ensuring a successful and efficient migration, as it outlines the key objectives, strategies, and milestones for your project while also considering the potential impact on your business operations.

Developing a cloud transformation plan involves a systematic approach that takes into account your organization's unique requirements, industry-specific regulations, and the potential challenges that may arise during the migration process. A robust plan helps you allocate resources effectively, manage risks, and ensure that your cloud migration aligns with your organization's strategic goals.

In this section, we will provide practical examples and guidance on how to develop a comprehensive cloud transformation plan, from defining the project scope and objectives to planning for ongoing management and optimization. By the end of this section, you will have a solid understanding of the key components of a cloud transformation plan and how to tailor it to your organization's specific needs and goals.

Define project scope and objectives

Defining the project scope and objectives is a critical first step in developing your cloud transformation plan. By clearly outlining the scope, you can identify the specific systems, applications, and infrastructure components that will be migrated to the cloud. This clarity enables you to create a focused and effective migration plan that addresses your organization's unique needs.

Begin by conducting a thorough inventory of your current IT assets, including hardware, software, and network components. Identify which systems and applications are suitable for cloud migration, considering factors such as compatibility, performance, and potential cost savings. For example, you may decide to migrate a legacy application to take advantage of the scalability and flexibility offered by cloud-based infrastructure.

Next, establish measurable objectives for your cloud transformation project that align with your organization's goals. These objectives might include:

- **Cost savings:** Estimate the potential cost savings that can be achieved through reduced hardware investments, lower maintenance costs, and improved resource utilization in the cloud.
- **Performance improvements:** Identify the areas where migrating to the cloud can lead to better application performance, such as reduced latency, faster response times, or improved availability.
- **Increased security:** Determine how cloud-based security services, such as encryption, identity management, and threat detection, can help you enhance the overall security of your systems and data.
- **Greater scalability and flexibility:** Assess how migrating to the cloud can provide your organization with the ability to quickly scale resources up or down in response to changing business demands.

Defining the project scope and setting measurable objectives will allow you to ensure that your cloud transformation plan is focused on achieving the desired outcomes and aligns with your organization's strategic goals. This clarity will help you allocate resources effectively, monitor progress, and ultimately ensure a successful migration to the cloud.

Select the appropriate migration strategy

Selecting the appropriate migration strategy is a vital step in developing your cloud transformation plan. By choosing the most suitable strategy for your project, you can ensure a smooth and efficient migration that meets your organization's specific requirements and constraints, such as budget, timeline, and technical expertise.

There are several migration strategies to choose from, including lift-and-shift, replatforming, and refactoring. Each strategy has its advantages and considerations, and understanding them can help you determine the best approach for your cloud transformation project:

- **Lift-and-shift:** This strategy involves moving your existing systems and applications to the cloud with minimal modifications. The lift-and-shift approach is typically faster and more cost-effective than other strategies, as it requires fewer changes to your existing infrastructure. However, it may not fully leverage the benefits of cloud-native features and may limit potential cost savings or performance improvements.

- **Replatforming:** Replatforming involves making minor modifications to your applications or infrastructure to better adapt them to the cloud environment. This strategy can help you take advantage of some cloud-native features, such as auto-scaling or managed services, while still minimizing the time and effort required for migration.

- **Refactoring:** Refactoring involves rearchitecting your applications or systems to fully leverage the capabilities of cloud-native technologies. This strategy can deliver significant benefits in terms of performance, scalability, and cost savings but may require more time, effort, and technical expertise to implement.

To select the most appropriate migration strategy for your cloud transformation project, consider the following factors:

- **Budget:** Assess the costs associated with each migration strategy, including development, testing, and infrastructure expenses. Choose a strategy that aligns with your organization's financial constraints and provides the best return on investment.
- **Timeline:** Determine the time required for each migration strategy, taking into account planning, development, testing, and optimization efforts. Select a strategy that meets your project deadlines and minimizes disruption to your business operations.
- **Technical expertise:** Evaluate your organization's technical capabilities and the level of expertise required for each migration strategy. Choose a strategy that aligns with your team's skills and experience, or consider investing in training or hiring external experts if necessary.

Selecting the appropriate migration strategy for your cloud transformation project will allow you to ensure a smooth and efficient transition to the cloud that meets your organization's unique requirements and constraints.

Develop a detailed project timeline

Developing a detailed project timeline is an essential step in creating a successful cloud transformation plan. By outlining the key milestones and deadlines, you can effectively allocate resources, monitor progress, and ensure that all aspects of the project are completed within the desired timeframe.

To create a comprehensive project timeline for your cloud transformation project, follow these steps:

- **Break down the project into smaller tasks:** Identify the specific tasks required for each phase of the project, such as planning, migration, testing, and optimization. This will help you gain a clear understanding of the scope and complexity of the project and enable you to allocate resources effectively.

- **Estimate the duration of each task:** Determine the time required to complete each task, taking into account factors such as technical complexity, resource availability, and potential dependencies. Be realistic in your estimations and consider any risks or challenges that may impact the project schedule.
- **Set key milestones and deadlines:** Define the critical milestones and deadlines for your project, such as the completion of planning, the start of the migration, and the go-live date. These milestones will help you track progress and ensure that the project remains on schedule.
- **Allocate sufficient time for each phase:** Ensure that you allocate enough time for each phase of the project, including planning, migration, testing, and optimization. This will help you avoid potential bottlenecks and ensure that all tasks are completed within the desired timeframe.
- **Incorporate buffer time:** Include buffer time in your project timeline to account for unexpected delays or issues that may arise during the migration process. This will help you maintain flexibility in your schedule and minimize the risk of missing critical deadlines.
- **Continuously monitor and adjust the timeline:** Regularly review your project timeline and make any necessary adjustments based on the actual progress, challenges, or changes in priorities. This will help you maintain control over the project and ensure a successful and efficient transition to the cloud.

Developing a detailed project timeline for your cloud transformation project will allow you effectively manage resources, monitor progress, and ensure that all milestones and deadlines are met. This will ultimately contribute to a successful and efficient migration to the cloud that aligns with your organization's goals and objectives.

Establish a governance framework

Establishing a governance framework is a crucial component of your cloud transformation plan. By implementing a well-defined framework, you can ensure that your project adheres to organizational policies, industry-specific regulations, and best practices. This will help you manage risks, maintain compliance, and optimize costs throughout the migration process.

To establish a governance framework for your cloud transformation project, consider the following steps:

- **Define clear policies and guidelines:** Develop a set of policies and guidelines that outline the standards, processes, and best practices for your cloud transformation project. These policies should address key aspects such as security, compliance, risk management, and cost optimization. Make sure to align your policies with your organization's strategic goals and industry-specific regulations.

- **Assign roles and responsibilities:** Clearly define the roles and responsibilities of your team members, ensuring that each person has the necessary authority and accountability to manage their respective areas of the project. This will help you maintain control over the project and ensure that all tasks are completed efficiently and effectively.

- **Implement security and compliance measures:** Establish processes for managing security and compliance throughout the cloud transformation project, such as encryption, identity and access management, and data protection. Ensure that your chosen cloud provider meets the necessary security standards and compliance requirements for your industry.

- **Develop a risk management process:** Create a process for identifying, assessing, and mitigating risks associated with your cloud transformation project. This should include regular risk assessments, clear escalation paths, and contingency plans to address potential issues that may arise during the migration process.
- **Monitor and optimize costs:** Implement cost management processes to monitor and optimize expenses throughout the project. This may include tracking resource usage, setting budget limits, and identifying opportunities for cost savings or efficiencies.
- **Regularly review and update the governance framework:** Continuously evaluate the effectiveness of your governance framework and make any necessary adjustments based on changes in your organization's objectives, industry regulations, or best practices. This will help you maintain a robust and effective governance framework that supports your cloud transformation project.

Establishing a governance framework for your cloud transformation project will allow you to ensure that your migration adheres to organizational policies, industry-specific regulations, and best practices. This will help you manage risks, maintain compliance, and optimize costs, ultimately contributing to a successful and efficient transition to the cloud.

Plan for ongoing management and optimization

Planning for ongoing management and optimization is an essential aspect of your cloud transformation project. By developing a comprehensive plan for the continuous monitoring, performance tuning, and cost management of your cloud-based infrastructure, you can ensure that your organization fully leverages the benefits of the cloud and maintains optimal performance and cost efficiency.

To develop a plan for ongoing management and optimization of your cloud-based infrastructure, consider the following steps:

- **Define roles and responsibilities:** Clearly outline the roles and responsibilities of your team members for managing and optimizing your cloud infrastructure. This should include tasks such as monitoring resource usage, identifying performance bottlenecks, and implementing cost-saving measures.
- **Establish monitoring and performance tuning processes:** Implement processes for continuously monitoring the performance of your cloud-based systems and applications, using tools and services provided by your cloud provider or third-party solutions. Develop a strategy for identifying and addressing performance bottlenecks, such as optimizing resource allocation, leveraging auto-scaling, or fine-tuning application code.
- **Implement cost management processes:** Establish processes for managing and optimizing costs in your cloud environment, including setting budget limits, tracking resource usage, and identifying opportunities for cost savings. Leverage cost management tools and services offered by your cloud provider to gain insights into your spending patterns and identify areas for improvement.
- **Create a training and skill development plan:** Assess the current skill set of your team members and identify any gaps in knowledge or expertise required for the ongoing management and optimization of your cloud infrastructure. Develop a plan for addressing these gaps through training programs, certification courses, or hiring additional staff with the necessary skills.

- **Continuously evaluate and optimize your infrastructure:** Regularly review the performance, security, and cost efficiency of your cloud infrastructure, and make any necessary adjustments to ensure optimal operation. This may include upgrading to newer services or technologies, implementing best practices, or rearchitecting applications for better performance or cost savings.

- **Establish a feedback loop:** Encourage ongoing communication and collaboration among your team members to continuously identify areas for improvement, share best practices, and address any challenges that may arise in the management and optimization of your cloud infrastructure.

Planning for ongoing management and optimization of your cloud-based infrastructure will allow you to ensure that your organization fully leverages the benefits of the cloud while maintaining optimal performance and cost efficiency. This will help you maximize the return on your cloud transformation investment and support your organization's strategic goals.

Practical: Executing the Cloud Transformation Project

In section you will gain hands-on experience in carrying out the crucial steps involved in successfully transitioning your organization's infrastructure to the cloud. This practical exercise will guide you through the entire cloud transformation process, from setting up the cloud environment and migrating your data and applications, to testing, monitoring, and optimization.

Throughout this section, you will learn how to apply the best practices and strategies discussed in previous modules to real-world scenarios. You will also explore the importance of training and supporting your team members during the cloud transformation journey, ensuring they are equipped with the necessary skills and knowledge to manage and optimize the new cloud-based infrastructure effectively.

By the end of this practical exercise, you will have a comprehensive understanding of the entire cloud transformation process and be well-prepared to execute a successful cloud migration project for your organization.

Preparing the environment

As you begin executing your cloud transformation project, the first step is to prepare the environment by setting up and configuring the selected cloud platform. This involves creating and configuring the necessary resources, such as virtual networks, storage, and security settings, to ensure a smooth and successful migration.

Start by creating a virtual network within your chosen cloud platform. This will serve as the backbone for your cloud infrastructure, connecting your virtual machines and other cloud resources. Configure the network settings according to your organization's requirements, such as IP address ranges and subnet configurations.

Next, set up cloud storage services for your applications and data. Depending on your needs, you can choose from different storage types like object storage, block storage, or file storage. Ensure that you create and configure the storage resources to meet your performance, availability, and redundancy requirements.

Finally, establish security settings to protect your cloud infrastructure. This includes configuring identity and access management (IAM) policies to control access to your cloud resources, setting up firewall rules to restrict traffic to and from your virtual network, and implementing encryption for data at rest and in transit.

Carefully preparing your cloud environment will allow you lay a solid foundation for the migration process and create a secure and efficient infrastructure to support your organization's operations in the cloud.

Migrating data and applications

Once your cloud environment is prepared, it's time to migrate your data, applications, and systems from the on-premises infrastructure to the cloud. To achieve this, follow the migration strategy you've chosen, such as lift-and-shift, replatforming, or refactoring. Each strategy has its own advantages and challenges, so make sure you've selected the one that best suits your organization's needs and goals.

During the migration process, carefully plan and execute each step to minimize downtime and ensure a smooth transition. Depending on the complexity of your applications and infrastructure, you might choose to migrate components in phases or use a hybrid approach, gradually moving workloads to the cloud while maintaining some on-premises resources.

To migrate data, leverage the cloud provider's data transfer services or third-party tools that support high-speed data transfer and minimize the impact on your network bandwidth. Verify that all data has been transferred accurately and securely, maintaining data integrity and consistency throughout the process.

For applications, follow best practices for deploying and configuring them in the cloud environment, taking into account any necessary modifications for compatibility, performance, or security. Test your applications thoroughly during the migration to ensure they function correctly and meet your organization's performance and availability requirements.

Carefully planning and executing your data and application migration will allow you to minimize disruption to your organization's operations and successfully transition your workloads to the cloud.

Testing and validation

After migrating your data and applications to the cloud, it's essential to conduct thorough testing and validation to ensure everything functions correctly in the new environment. This step helps you identify and address any potential issues that may have arisen during the migration process.

Begin by testing the functionality of your applications, verifying that all features and components work as expected. This may involve running automated test suites, performing manual tests, or conducting user acceptance testing (UAT) with a group of users who can provide feedback on the system's performance and usability.

Next, assess the performance of your applications, ensuring they meet the required response times, throughput, and resource usage levels. Use monitoring tools to gather performance metrics and identify any bottlenecks or areas that require optimization.

Make necessary adjustments to configurations, scaling settings, or application code to improve performance.

Additionally, validate the security of your cloud environment, confirming that all access controls, encryption, and firewall settings are correctly configured. Conduct vulnerability assessments and penetration testing to identify and remediate potential security risks.

Finally, ensure that your cloud environment complies with any relevant industry regulations or standards, such as GDPR, HIPAA, or PCI DSS. Review your governance framework and verify that all necessary documentation, audits, and controls are in place to demonstrate compliance.

Rigorously testing and validating your cloud environment will allow you to have confidence that your migrated applications and systems function correctly, meet performance and security requirements, and comply with industry regulations.

Monitoring and optimization

Once your cloud environment is up and running, it's crucial to implement monitoring tools and techniques to keep track of the performance and usage of your cloud resources. Monitoring allows you to proactively identify and address any issues that may arise, ensuring optimal performance and cost efficiency.

Start by selecting monitoring tools that are compatible with your chosen cloud platform. Many cloud providers offer built-in monitoring services, such as Amazon CloudWatch, Azure Monitor, or Google Cloud Operations Suite. These services can be complemented by third-party tools, depending on your specific needs and preferences.

Configure your monitoring tools to collect metrics related to performance, availability, and resource usage. Set up custom alerts and notifications to inform you of any issues, such as high resource utilization, degraded performance, or security incidents. Regularly review these metrics to gain insights into the health of your cloud environment and identify areas for improvement.

Based on the data collected, continuously optimize your cloud environment to ensure cost efficiency and maintain high-performance levels. This may involve adjusting resource allocation, scaling settings, or application configurations. Additionally, keep an eye on your cloud spending and make use of cost optimization techniques, such as selecting the right instance types, using reserved instances or savings plans, and eliminating unused or underutilized resources.

Actively monitoring and optimizing your cloud environment will allow you to ensure that your infrastructure remains cost-effective and maintains the performance levels needed to support your organization's operations.

Training and support

As you transition to a cloud-based infrastructure, it's essential to provide training and support to your team members to help them adapt to the new environment and processes. Equipping your team with the necessary skills and knowledge is critical for effectively managing and optimizing your cloud resources.

Start by identifying the training needs of your team members, considering their current skill sets and the specific requirements of your cloud environment. This may include training on cloud platform fundamentals, specific services or technologies, security best practices, or cost management techniques.

Select appropriate training resources and methods to meet your team's needs. This can include online courses, workshops, webinars, or hands-on lab exercises. Many cloud providers offer comprehensive training programs and certifications, such as AWS Training and Certification, Microsoft Learn, or Google Cloud

Training, which can help your team build the skills they need to succeed in the cloud.

In addition to formal training, create a culture of continuous learning and improvement within your organization. Encourage team members to share their knowledge, experiences, and best practices with one another, fostering collaboration and growth. Provide ongoing support, such as mentoring, coaching, or access to expert resources, to help your team overcome any challenges they may encounter as they work with the cloud environment.

Investing in the training and support of your team members will allow you to ensure they are well-prepared to manage and optimize your cloud-based infrastructure, ultimately contributing to the success of your cloud transformation project.

Module 10 Exam Questions

1. Which of the following is NOT a common cloud migration strategy?

a) Lift-and-shift

b) Replatforming

c) Refactoring

d) Dematerializing

2. Which cloud platform is known for its extensive range of services and global infrastructure?

a) Amazon Web Services (AWS)

b) Microsoft Azure

c) Google Cloud Platform (GCP)

d) IBM Cloud

3. What is the primary goal of a cloud transformation project?

a) To move all IT infrastructure to the cloud

b) To leverage the benefits of cloud computing

c) To replace all physical servers with virtual machines

d) To minimize the use of on-premises data centers

4. Which aspect of cloud migration involves minimal modifications to existing systems and applications?

a) Lift-and-shift

b) Replatforming

c) Refactoring

d) Cloud-native development

5. What is an essential component of a cloud transformation plan?

a) A detailed project timeline

b) A list of all cloud providers

c) A complete redesign of all applications

d) Outsourcing all IT operations

6. Which industry-specific regulation impacts healthcare organizations during cloud transformation?

a) HIPAA

b) GDPR

c) PCI DSS

d) FISMA

7. What is the primary purpose of a governance framework in a cloud transformation project?

a) To enforce strict coding standards

b) To ensure adherence to policies and regulations

c) To minimize the use of open-source software

d) To reduce the number of software licenses

8. Which cloud migration strategy is most suitable for taking advantage of cloud-native features?

a) Lift-and-shift

b) Replatforming

c) Refactoring

d) Cloud-native development

9. What is the primary focus of cloud cost optimization?

a) Reducing the number of cloud services

b) Minimizing resource usage and costs

c) Eliminating all cloud-based expenses

d) Reducing the number of data centers

10. What should be included in a cloud transformation project timeline?

a) Key milestones and deadlines

b) A list of all project team members

c) A complete inventory of all IT assets

d) A detailed budget for the project

11. In the context of cloud migration, what does "refactoring" mean?

a) Moving applications to the cloud without modifications

b) Making minor modifications to applications for better cloud adaptation

c) Rebuilding applications to take full advantage of cloud-native technologies

d) Removing all on-premises servers

Which factor should be considered when selecting a cloud 12. provider?

a) Number of available data centers

b) Organization's requirements and budget

c) Number of existing clients

d) Marketing strategies

13. What is the main benefit of a lift-and-shift migration strategy?

a) Cost reduction

b) Faster migration process

c) Improved performance

d) Enhanced security

14. What is an important aspect of ongoing cloud management and optimization?

a) Performance tuning

b) Infrastructure expansion

c) Application redesign

d) Staff reduction

15. Which of the following should be included in a governance framework?

a) Security and compliance measures

b) Hardware procurement processes

c) Software development methodologies

d) Annual performance reviews

16. What is a key consideration when evaluating the potential impact of cloud transformation on business operations?

a) Potential downtime

b) Cloud provider's reputation

c) Availability of open-source software

d) Number of available cloud services

17. What is the primary goal of a risk management process in a cloud transformation project?

a) To eliminate all potential risks

b) To identify, assess, and mitigate risks

c) To focus solely on security risks

d) To reduce the likelihood

18. What should be considered when developing a training and skill development plan for a cloud transformation project?

a) Employee preferences

b) Addressing knowledge gaps

c) The number of available training courses

d) The popularity of cloud platforms

19. Which aspect of a cloud transformation plan focuses on ensuring adherence to organizational policies and industry regulations?

a) Governance framework

b) Project timeline

c) Migration strategy

d) Cost management processes

20. When planning for ongoing management and optimization of cloud-based infrastructure, what should be outlined for team members?

a) Their roles and responsibilities

b) The number of cloud services to use

c) The expected increase in revenue

d) The amount of cloud storage needed

Answers:

1. d
2. a
3. b
4. a
5. a
6. a
7. b
8. c
9. b
10. a
11. c
12. b
13. b
14. a
15. a
16. a
17. b
18. b
19. a
20. a

Conclusions and Next Steps

As we reach the end of this book, it is clear that the cloud transformation revolution is upon us, and its impact on our professional lives and businesses is undeniable. Throughout this journey, we have debunked the myth of cloud transformation as a mere trend, explored its potential to drive efficiency and innovation, and discussed the security and organizational implications of adopting cloud technologies. We have also delved into the strategies for cultivating a flexible mindset, navigating the complex landscape, and developing cloud leadership and organizational readiness.

The future of business will be shaped by the seamless integration of cloud technologies with human ingenuity, collaboration, and adaptability. By understanding the strengths and limitations of cloud transformation, we can harness its power to create a sustainable and agile future for all. However, to truly make a difference, it is essential for each individual to take proactive steps towards embracing the changes brought by the cloud transformation era.

As a game changer, you have the power to make a lasting impact on your career and the world. Here are the next steps you can take to embrace the cloud transformation and create positive change:

1. **Enroll the online program at TheBestMBA.org:** Certify your skills, gain in-depth knowledge, practical tools, and valuable insights from our comprehensive curriculum. Our program is designed to help you thrive in the cloud transformation era and adapt to the ever-changing demands of the workforce.
2. **Engage in continuous learning:** Stay ahead of the curve by constantly updating your skills and knowledge. Embrace lifelong learning and take advantage of the vast resources available online, such as courses, webinars, podcasts, and articles.

3. **Expand your network:** Connect with industry leaders, experts, and fellow game changers through events, conferences, and workshops. Foster meaningful relationships and collaborations that will propel you forward in your career.
4. **Seek mentorship and guidance:** Reach out to experienced professionals in the field of cloud and related industries to learn from their insights and experiences. Their wisdom can help you navigate the complexities of the cloud landscape and ensure your success.
5. **Volunteer or collaborate on cloud projects:** Gain hands-on experience and demonstrate your capabilities by working on real-world cloud projects. This not only enhances your skills but also showcases your value to potential employers and collaborators.

By taking these steps, you will be well on your way to becoming an influential game changer in the cloud era. Remember, the journey towards a sustainable and inclusive future begins with each individual's commitment to adapt, learn, and grow. Embrace the possibilities that cloud transformation offers and let your unique human skills shine through as you create a better tomorrow.

The future is now. Are you ready to shape it?

Join the IIEC Cloud Certification Program today available at https://thebestmba.org and become a game changer.

I hope you enjoyed this journey together, and as I mentioned at the beginning of this book, I would love for you to share your experience with me after completing this program. Feel free to message me via LinkedIn.

https://www.linkedin.com/in/angelberniz

To your success,

Angel Berniz

Made in the USA
Monee, IL
04 November 2026

67384350R00233